Build
Multi

9780750628556

D1652122

Build Your Own Multimedia PC

A complete DIY guide to renovating and constructing personal computers

IAN SINCLAIR

NEWNES

Newnes
An imprint of Butterworth-Heinemann
Linacre House, Jordan Hill, Oxford OX2 8DP
A division of Reed Educational and Professional Publishing Ltd

A member of the Reed Elsevier plc group

OXFORD   BOSTON   JOHANNESBURG
MELBOURNE   NEW DELHI   SINGAPORE

First published 1996
Reprinted 1997

© Ian Sinclair 1996

British Library Cataloguing in Publication Data
A catalogue record for this book is available from the British Library

ISBN 0 7506 2855 3

Cover photograph supplied courtesy of, and © copyright 1993, Maplin Electronics plc,
all rights reserved. Availability and specification of products shown is subject to change
without notice.

Maplin Electronics plc. supply a wide range of electronic components and other products
to private individuals and trade customers. Tel: (01702) 552911 or write to Maplin
Electronics, PO Box 3, Rayleigh, Essex, SS6 8LR for further details of product catalogue
and location of regional stores.

Typeset by Butford Technical Publishing, Bodenham, Hereford
Printed in Great Britain by Hartnolls Limited, Bodmin, Cornwall

Contents

Acknowledgements

Photographs used in this book are supplied courtesy of and copyright of Maplin Electronics plc, all rights reserved. Availability and specification of products shown is subject to change without notice. Maplin Electronics plc implies no warranty or endorsement with regard to information presented in this book. Save insofar as prohibited by English law, liability of every kind, including negligence, is disclaimed as regards any person thereof.

Maplin Electronics Plc supply a wide range of computer hardware, computer accessories, electronic components and other products to private individuals and trade customers. Telephone (01702) 552911 or write to Maplin Electronics, PO Box 3, Rayleigh, Essex SS6 8LR for further details of product catalogue and the location of regional stores. The Maplin catalogue is also available at leading bookshops.

Preface

The mass production of PC machines of all types, and the legal actions in the USA that have allowed manufacturers other than Intel to produce chips that can be described as 80386 or 80486, have led to a thriving assembly industry in PC machines, so that it is now possible for anyone with facilities for assembling circuit boards into cases to put together PCs with capabilities better than the older IBM AT style of machine and equal to all but a few modern designs. The sheer number of small-scale suppliers, and the standardization of design, indicates how easy this work can be, using plug-in boards from the lowest-cost sources. Construction, in this sense, can mean assembly, and not necessarily much assembly in some examples.

The small-scale assemblers, some of whom are offering a stripped-down 486 machine with 4Mb of memory for £500 or less, cannot offer much more than hardware. They cannot offer a manual that makes much sense to the first-time user, and even an experienced PC user can be baffled by a new machine if little or no information is available. In particular, if Amstrad users are moving from the old 1512 and 1640 models to modern clones, they will find their Amstrad experience of little use because the Amstrad is mechanically incompatible, particularly in the relationship of power supply and monitor to the main board of the computer.

The title of this book suggests that multimedia use is the aim, but any reasonably competent PC will allow you to upgrade to multimedia, and the point of using the word in the title is that you should aim at this standard even if you have no immediate intention of using multimedia or equipping for it. In fact, what a PC is used for is very much an individual matter and this book is aimed at all shades of taste. You might, for example, require only the basics of word processing, so that it would be gross overkill to use a machine that featured the fastest processor, with multimedia capabilities and huge resources of memory. Sections I and II show just how cheaply and easily a machine which was regarded as 'state-of-the-art' in 1992 can be put together, and such a machine is still as much as 75% of computer users (or prospective users) really need.

You might, however, want to go further. Though multimedia extensions (meaning a sound card, loudspeakers and CD-ROM drive) can be added to the older type of 386 machine, you need something faster and with more memory for better performance. If, in addition, you want to be able to use more recent programs like Microsoft Office and, in particular, you want to be able to run Windows 95, then you need the fastest machine you can afford, with as much memory and hard drive space as you can afford. A reasonable bottom limit is a 486/66 machine (meaning 80486 processor and 66 MHz clock speed) with 8 Mbyte of memory and a 512 Mbyte hard drive. Section V is devoted to this type of upgrade or initial buy. The idea of *buying* a machine of this calibre to start with might seem rather odd for a book with this title, but the point is that in a year or so you might need to upgrade again, and if you did not buy wisely you cannot upgrade easily.

This book is a form of manual that will cover the construction of a multimedia PC, either from scratch or following the more common method of buying a low-cost machine from a local assembler, or from other sources such as auctions, and improving it as required. This book will also be a useful reference text for users of all the machines that can be described as 'generic', machines which are very closely compatible with the old IBM AT design but with enhanced facilities. If your low-cost PC comes provided with a manual that can be politely described as rudimentary, this book is for you. Since it is assumed that you are building a new PC or updating an old one, information that is relevant only to old machines is omitted unless you need to know it for the purposes of updating. The use of old machines (prior to the 80386 models) is a problem that makes many books on the PC machine bulkier than they need be just to account for all the variations that are required on older machines.

One point that often worries prospective DIY builders is that their machine will be non-standard. The fact is that a home-constructed machine is likely to be totally standard, more so than some big-name varieties, and more adaptable to upgrading. Another worry is that some inadvertent action will destroy the whole machine, and this too is a myth unless you make a habit of dropping hammers into equipment. Perhaps we should add also the worry that the machine will be damaged in some way by unsuitable software or when a program locks up – as this book points out, the computer clears its memory when it is switched off, and a fault in a program cannot affect any other program that is run after re-starting like this.

As it happens, building a PC from scratch is often more expensive than buying a machine from some of the small-scale firms, and most private owners take the course of buying only as much as they need of an assembled machine – often a case, PSU and motherboard only. Many suppliers specialise in this type of 'bare-bones' machine, and because the parts are usually standardized, such machines are easy to work with and to upgrade. The point of assembling your PC in this way is that you can also upgrade for yourself, avoiding the high costs that are so often associated with changing hardware. A second-hand computer, after all, depreciates even faster than a second-hand car.

Another route which is now significant is to buy machines that have been discarded by local authorities and other corporate users – the more a local authority or nationalized service complains about lack of money the more computers they appear to scrap (not sell), simply because they are not the most recent models. These machines are found at auctions and at car-boot sales, often at very low prices. Some are older 80386 types, others are almost new 80486 machines which have been used in networks and which may lack a hard drive. Prices of £25 to £75 make this a very encouraging start and one which is much cheaper than buying all parts separately.

The aim of the book is to provide information for anyone taking any of these routes, because no manual will be available. Since many readers of this book are likely to be experienced in electronics, some aspects of computer circuitry and disk recording are explained in more detail than would be relevant to the reader with no electronics background. Other than these paragraphs, the book is intended to be used by newcomers and experienced users alike, either in computing or in electronics.

Finally, this book is also intended as a reference for anyone who is updating a machine, either a minor update such as replacing a disk drive or a major update such as replacing a motherboard. Since the effects of construction and upgrading cannot be judged without the essential software, the bare essentials of using MS-DOS and Windows are also included, along with a chapter on printers. Note, however, that you cannot really use multimedia unless you use Windows.

Ian Sinclair
May 1996

Section I

Mainly for masochists – assembly from scratch

Chapter 1

Preliminaries, fundamentals and buying guide

This chapter is intended for the prospective assembler of a PC whose experience has been in electronics construction rather than in computing. If you are already well-experienced in computing and want to experience the joys of self-assembly then read this by all means, but be prepared to skip some explanations that are intended for the newcomer to computers.

To start with, the type of machine that we now describe as a PC means one that is modelled on the IBM PC type of machine that first appeared in 1980. The reason that this type of machine has become dominant is the simple one of continuity – programs that will work on the original IBM PC machine will work on later versions and will still work on today's machines. By maintaining compatibility, the designers have ensured that when you change computer, keeping to a PC type of machine, you do not need to change software (programs). Since the value of your software is much greater than the value of the hardware (the computer itself), this ensured that the PC type of machine became dominant in business and other serious applications. Other machines are not compatible with the PC or with each other, have less choice of software and more expensive components.

Compatibility works only one way however, and software that is being written now will not necessarily run on old machines, though a lot of programs still do. The main benefit of a long-established design is that components are remarkably cheap and reliable, and that the layout of machines is more or less standardized. Though you can build an old-style machine for a low price, it is well worth the small amount of extra cash to construct or buy a PC machine that is reasonably up to date in design, because this allows you the luxury of being able to use any software written for the PC, not just the older programs. You should aim for a machine that can run the modern Windows system (see Chapter 9).

In short, a real PC machine is currently identified by the following points:

1   It uses a microprocessor which is of the Intel type, with a type number that starts with 80 and ends in 86 – the 8086 and 80286 are very early types, and you should go for the 80386 or 80486. The new Pentium chip is also of this family.

2   It uses a program called MS-DOS or PC-DOS as a master controlling system (an operating system), to enable it to load and run all other programs.

What and why?

Before you think of spending any hard-earned money on a PC machine you need to think carefully about what your needs are. Generally, when you buy a computer for the first time you have some main use in mind. This might be word-processing because you need to write reports, articles, sermons, notes, books or whatever. It might be database use, because you need to keep track of several thousand items in a mail-order catalogue or points in a sports league or references in newspapers. It might be a spreadsheet because you need to keep tables of items in a way that allows you to work out totals and averages, or it might be a book-keeping program for your business needs.

Whatever your needs are initially, once you have experienced the advantages of working with the computer, and adapted your methods to the use of the computer, you will want to make it work harder for you. You are likely to buy other main items of software, and you are also likely to want to use the programs that are collectively called *utilities*.

The point about adapting your methods is important, incidentally. Any task that you have previously done by hand usually needs to be done quite differently by computer – one of the few exceptions is book-keeping. The computer forces you to work in a different way, but as a compensation it allows you to work with greater freedom. You can make corrections and alterations easily, for example. Try typing an article and then inserting a 20-word amendment in the middle of the work. This is simple routine stuff when you use a word-processor, tedious and awkward when you use a steam typewriter. Try using a card index to produce a list of all

UNF-threaded bolts in size 6 with cadmium plating and hex heads – it's easy with the computer running a database, but you must have organized the information correctly in the first place, and not as you would for a card index. When users feel disappointed with the use of a computer, the reason is almost always that they are trying to make the machine work in the way that they formerly worked with pen and paper.

Whatever you bought the machine for in the first place, you are likely to find that you have many more applications for it after a year or so. This is when you may come up against restrictions that seemed unlikely when you first bought the machine. You may need more memory, to run larger programs, more disk space, faster actions, a better monitor. If you chose wisely initially you should find that your machine is capable as it is, and even if you went for the minimum that you could get away with, wise planning will ensure that you can easily upgrade the machine to do what you want. That sort of action is also covered in this book. Remember, however, that upgrading a computer is rather like upgrading a Hi-Fi system – it can be continued forever and eventually the gains are too small to notice. You have to ask yourself continually if an upgrade really fulfils a need or whether it simply allows you to use a more elaborate version of a program that serves you equally well at the moment.

Upgrading the machine is not confined to simply increasing its memory and ability to deal with more complex programs. Add-on boards exist for virtually every purpose for which a computer can be adapted, and the PC machine forms an excellent basis for experimental work for anyone with experience in electronics. One such add-on that is gaining popularity is the use of multimedia, with a compact disc player and a sound system interfaced to the machine. This allows for programs that incorporate text, pictures and sound to be replayed, using the compact disc as the storage medium because of its large storage capacity of around 600 Mbyte.

A less-trumpeted aspect is control engineering, using analogue-digital converter cards, allowing the PC to act as part of a control system for process engineering, environmental control and so on. Similar add-on cards can also be used to make the PC part of a security system with the advantage that the response can be altered by programming the machine for yourself. You can also couple in devices such as bar-code readers and printers to make the PC part of a data system. All of these actions are too specialized for this book, but you should be aware that they exist and if you are interested, look out for books that deal with these topics. Chapter 11 covers the bare bones of such additions.

The components

A basic PC system consists of a main casing that contains the mother-board and disk drives, along with a separate keyboard and monitor. The keyboard and the contents of the main casing are the components that lend themselves to DIY assembly, and the monitor is bought as a single, separate, item. You can use either a monochrome or a colour monitor and for many purposes the monochrome monitor is superior in clarity, as well as being much cheaper.

The main advice here is to avoid working with the older components. The 8088, 8086 and 80286 microprocessors are now obsolete and though there are millions of PCs working happily with these processors, they are not capable of running the mainstream of modern programs. In particular, if, after reading Chapter 9, you decide to use the Windows system, you need a machine which is initially as capable as you can afford, and which can easily be upgraded, particularly with more memory, later.

If you are happy using older programs that are not too demanding of the machine, the earlier part of this book is for you. Parts for 80386 computers are now very cheap and easy to obtain, as are some discontinued models. If, however, you want to run Windows 95 and all that goes with it, the minimum standard to aim at is an 80486 machine and preferably a Pentium type, and this is covered in more detail in Chapter 12.

The assembly of a PC machine from scratch is, if anything, easier than making a working model from old-style Meccano, with the difference that you do not start with a full kit of parts. The comparison is not entirely fanciful, either, because a PC is put together using bolts of standard types, and circuit boards that plug into position; no elaborate tools are required nor vast experience needed. What you need to know is what parts you need, where to buy them and how to put them together. You do not need to know how to solder, and the highest order of electrical work you will be called on to do will be to connect up a standard mains plug.

The tools you need are mainly screwdrivers, preferably in the smaller sizes, and both plain and cross-head types. A pair of pliers is also useful though seldom essential. Other than these you need common sense (square plugs do not go into round holes) and some motivation (such as lack of money or fascination with computers). One useful point about assembling your own PC is that you can do it step by step. If cash is limited, you can buy one part each month until you complete the assembly.

On the subject of cost, assembling a PC from scratch is always going to be more costly than buying a new machine made from the same parts and bought from the cheapest source after some shopping around. The lowest-cost suppliers are almost always going to be small-scale mail-order suppliers and because they work on lower margins than the others they are more vulnerable to problems like slow payers, worried banks and strikes in delivery services. Because of this, you should always assume that such a supplier is unlikely to be a permanent fixture, and you should not part with real money. If you pay using a credit card (either in person or over the telephone) you have the protection of the credit card company. If the supplier vanishes overnight you will not lose because your card account will not be debited. There is no other way of paying that is so secure.

Looking for suppliers is easy if you subscribe to magazines such as PC Plus or PC Answers, whose advertisers contain several that specialize in parts for the DIY assembler. Do not confine yourself to these, however, because shopping around is important, and you may find bargains from suppliers who make no claims to cater for the assembler but who nevertheless hold an immense stock of PC parts at low prices. A few hours spent with these magazines can save you a lot of hard-earned cash. In particular, if you have a background in electronics and have a Maplin catalogue, you can use this single source for all your computer components, with the advantage of dealing with a well-known, soundly established, and reliable supplier.

All of this refers to standard desktop machines. Portable computers are quite another thing because there is no standard design, the parts are difficult to obtain and costly, and you need a ten year apprenticeship as a sardine-packer to be able to work on them. There are many buyers of portable machines, but fewer serious users whose needs would not be as well served with a notebook and a pencil.

The essential bits

The essential main bits of a PC are the casing (with power supply), the motherboard, and the disk drive. To check that a machine is working you also need a monitor, but since this is bought ready-made we do not count it as part of a DIY project. Do not attempt to convert a TV receiver into a monitor, or convert an old monitor into something suited to a modern PC, unless you have considerable experience of working on TV equipment (note in particular that any monitor must use an earthed chassis) and you have suitable circuit diagrams to work from. The reasons for this are noted in Chapter 4. Monitors from other types of machines will not necessarily be suited to a PC computer, though a few are adaptable. Be particularly careful of monitors, particularly large-screen monitors, offered as bargains. Some of these work with non-standard graphics boards which must be supplied along with the monitor, because they cannot be connected to a VGA board, but you cannot be sure that your software will allow the monitor to be used correctly, if at all.

With the essential bits in hand you can connect up a working PC machine, though it will not necessarily do everything that you want. From that stage, however, you can add other facilities by plugging in additional circuit boards, called cards, to extend the capabilities of the machine. You can also plug in additional memory units (called SIMMs), because whatever you do with a machine is likely to require more memory sooner or later, unless you start with as much memory as the motherboard can take. At the time of writing, memory is expensive but the price will inevitably come down again, so don't buy until you really need to.

The motherboard, as the name suggests, is the main printed circuit board of any PC machine which carries any other boards (or cards) that are added. It is a multi-layer board, and you must never drill it or cut it because the tracks that you see on it are only the surface tracks. The motherboard contains the main microprocessor chip (the CPU), and the type of CPU that is used determines the performance of the computer. At the time of writing, low-cost motherboards used the Intel 80386SX or an equivalent made by other suppliers such as AMD. These chips are graded by number, and you should not consider using a motherboard with a CPU whose number indicates an earlier design, such as the 80286, 8086 or 8088. The 80286 was a fine chip in its day, but that day was in 1982, and the 8086 and 8088 are even older. At the time of writing, the most advanced chip available on motherboards was the Pentium, but prices were still high (£300 to £1000 for the motherboard), and 80486 motherboards, even in the fastest (100 MHz) version, were much lower.

The successor to the 80486 is called Pentium because a name can be patented, and a number can not. No supplier other than Intel can therefore offer a Pentium without incurring legal action, but you will find motherboards with chips described as 80586 which may (or may not) be compatible with the Pentium. A successor to the Pentium, codenamed P6, is already being tested – nothing ever remains steady in this business. Whatever may happen in the future, it is likely that if you have not felt a pressing need to have a computer up to now you probably don't need a computer that is at the forefront of technology.

The 80386 is capable of much more than 99% of what its users require, and though it is now regarded as older technology this is the one to go for if you want low-cost computing because it is readily available, low-priced, reliable, and likely to be around for a considerable time (after all, the 8088 was being supplied in new machines for twelve years).

As well as the main CPU and the sockets for memory SIMMs, the motherboard contains all the other supporting chips and the connections (or bus) between the CPU and other sections. The other notable feature of the motherboard is the provision of slots, sockets for cards that are plugged in to expand the use of the machine. A good motherboard should allow six or more of these slots, and these are considered in more detail in Chapter 2.

The point of all this is that a microprocessor is not a component that you can replace easily if you want to upgrade, even if the motherboard could take a different microprocessor and even if you could extract the old processor. When you want to upgrade a computer, the simple and sensible method is to replace the whole motherboard. You can transfer the memory SIMMs from the old board to the new one, and any slot-in cards as well. That way you take advantage of not only the new microprocessor, but improvements in motherboard design as well.

The casing contains the power supply for the PC, which is always the type referred to as switch-mode, along with space for the motherboard, a set of shelves, called bays, for disk drives, various LED indicators and switches, and a lot of empty space. The most useful type of casing is also the cheapest and uses a hinged lid. Do not be tempted by miniature cases or very tall tower-block types, because they are more expensive and often difficult to work with if you want to add more disk drives and cards.

The side of the casing contains an opening for the main switch of the power supply unit, and the back right-hand side of the casing also has a set of six or more openings, usually temporarily covered by metal strips. These openings are at the slot positions, and each time you expand the capabilities of the computer by adding a card, it is likely that one coverplate will have to be removed to allow a connector mounted on the card to project outside the casing. These metal strips are each located by a single screw, usually of the cross-head variety. Do not use the machine with strips removed unless there are connectors to replace them, because this will upset the fan-driven airflow inside the machine.

The front panel of the casing has cut-outs for disk drives, and also a panel of switches and LEDs. This panel also contains a primitive form of lock which disables the keyboard, but since every computer user has at least one key that fits the lock it is not particularly useful. In any case, the wiring to the lock is accessible and easily changed.

The disk drive is the other essential, because a computer by itself is as useless as a CD player with no CDs. In computer jargon, the hardware is useless without software. Most of the memory of a computer is the kind described as volatile, meaning that it is wiped clear each time the machine is switched *off*, so that all the instruction codes that the machine needs to do anything have to be stored in a more permanent form. The two most familiar permanent forms are as a chip (a ROM or read-only memory chip) and as a magnetized disk. All desktop machines use both, and your motherboard will contain one or more ROM chips that contain a comparatively small amount of code. This is sufficient only to allow the machine to respond to the keyboard (in a limited way) and to operate the disk drive(s), also in a limited way. The rest of the essential codes, the operating system, are read in (loaded) from a disk.

The aim of this two-part storage is to allow the machine enough permanent instructions to read in an operating system that you can choose for yourself. The operating system is something that needs to be upgraded each time the capabilities of the computer are extended and if it were in ROM form it would require the ROM chips to be replaced. This action of using a small section of code to read in the rest of the operating codes is called bootstrapping (from the old myth of lifting yourself by your bootstraps) or *booting*, and the action of switching on a computer is referred to as booting up. The smallest portable machines do not use a disk drive, and they keep all of their operating system code in a ROM.

The main ROM chip is called the BIOS, meaning Basic Input Output Services, and this is a good description of what it provides. BIOS chips can come from a variety of suppliers, and the type that you find on your motherboard determines what additional facilities you may be able to call on. In particular, the BIOS chip controls a small CMOS RAM chip which is used to store machine information, using a battery backup so that the information is held permanently (as long as the battery lasts) but can be changed at will.

The disk drive is vital to a desktop machine, and all modern machines need at least two – one floppy drive which uses replaceable disks, and one hard drive which uses a set of disks that are fixed and encased in a sealed container. The floppy drive is used so that you can copy programs (software) that you buy and place on the hard drive, and for holding your own data. A hard drive has a limited life, so that it is essential to have a copy on floppy disks of everything on the hard drive. At one time it was possible to use a computer with a floppy drive only, but modern programs are too large to fit on a floppy drive, and the drive itself is too slow to allow a program to be run using just the floppy drive.

The assembly of casing, with power supply, the motherboard loaded with memory, and the disk drives needs only two cards to be plugged in and connected to make a working computer. One is the disk interface card which converts the numbers stored in the memory into pulses that can be recorded magnetically, and vice versa. The type of disk interface card that is almost universally used at the time of writing is the IDE card, the initials meaning Integrated Drive Electronics. Most IDE disk drive cards also carry out other tasks, so that they will add a printer port (connection) and a serial port (often two serial ports) to your basic computer as well.

The other card is the video graphics card which converts the computer pulses into video signals that a monitor can use. Once again there is an almost universal standard called VGA, and adding this card will allow you to connect up the monitor and see what happens when the computer is switched on. What happens from that point onwards depends on the software, which is the subject of Section III of this book. Because changing a graphics card is often an upgrading action, it has been placed in Section II.

Finally in this introduction, computing has enriched the English language with a large number of new words and new uses of old words. If these are new to you, Appendix A contains a glossary with full explanations, and some have been explained already in this chapter. Note that the word *disk* is used to refer to a magnetic computer disc, and the more familiar *disc* is used for CDs. This distinction has become important now that the CD format is used to distribute software.

Chapter **2**

Case, motherboard and keyboard

The casing of a computer is its most obvious hardware aspect, often labelled as boring by people who should know better (if you want bright transfers you can apply them for yourself). The full-scale desktop casing is 14" wide by 16" deep by 6.5" high, but there are many other varieties of various descriptions such as small-footprint, minicase and so on. The full size of case was needed in the early days when the main board (motherboard) was large and when a small disk drive implied a 5.25" full-height unit. Nowadays, slimmer casings can be used, and often are because motherboards are smaller and disk drives are slimmer. A few manufacturers use a tower-block form of casing in which the casing rests on its 14" x 6.5" side.

If you want to end up with a machine that uses the 80486 or Pentium chip and is equipped for multimedia, you should consider using a full-sized tower or mini-tower casing. These are more expensive than the desktop variety, but they allow for more to be added – for example, they usually have space for two hard drives as well as for a floppy, a CD-ROM drive and other items like a tape backup drive. Some of the desktop casings are not well-equipped for adding a large number of drives, and accessibility can be a problem with some designs. If you start by buying a low-cost machine, the tower version is often easier to work on.

The casing is not just something that should concern you if you are building a machine for yourself. The casing that a manufacturer has used for a machine determines quite critically what you can do in the way of adding new facilities and replacing the motherboard. Since a computer design can be considered as fairly new for only a few months, it is important to be able to upgrade a machine easily. Even if you are starting with buying a machine that is, by modern standards, obsolete, you should not neglect the case design, because it may make the difference between being able to upgrade cheaply by replacing a motherboard or expensively by replacing everything. The quality of the case is often the best way of deciding between two clones of unknown name and almost identical appearance. No matter how neat and tidy a small casing may look, it is not necessarily something that you will congratulate yourself about later.

What should you look for in a casing? A metal casing is important, because a metal casing greatly reduces the radiation of radio interference from the computer. Some big-name manufacturers in the past have used plastic cases and been obliged also to use metal sheets internally to comply with radio interference regulations. Using a metal case is a much easier solution, and the best form of case is the established flip-lid type which allows easy access to the interior simply by pressing two catches and lifting the lid up on its hinges. Plastic casings are not on offer for self-assembly, so if you are starting from scratch you will quite certainly be buying a steel casing. Figure 2.1 shows one of the Maplin range of casings of a type ideal for the constructor.

The overall size is not quite so important as the space for internal components, and one of the most important of these considerations is the number of disk drive bays. A disk drive bay, Figure 2.2, of the conventional sort is a shelf which is intended for the older type of 5¼" disk drive. Such a drive will slide into the bay from the front and be secured by screws at the side of the bay. The piece of casing at the front of the bay is usually a clip-on plastic portion which can be discarded when a floppy disk drive is fitted into the bay. A good case should offer at least

Figure 2.1 *A casing ideally suitable for home construction. This is the Maplin ZG64 type of casing in its full-size form.*

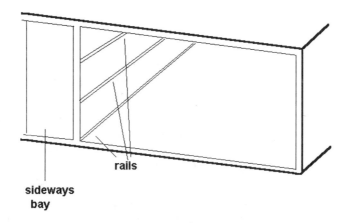

Figure 2.2 *A typical disk drive bay as seen from the front with the covering panel removed.*

three such bays, and preferably at least one other one mounted sideways and intended for a $3\frac{1}{2}$" drive – this will often be used for a hard disk drive but is sometimes used for the main floppy drive. The lowest of a set of three bays has a fixed front panel, and is intended for a hard drive.

The importance of drive bays cannot be over-estimated. You might think that with $5\frac{1}{4}$" drives almost obsolete there would be no point in using bays intended for this size, but developments in the last year or so have made their use essential, because many new enhancements to a PC system demand the use of such bays. If, for example, you want to use a CD drive, referred to as CD-ROM, now that these drives are more reasonably priced and serious software is being distributed in this format, you will find that installation requires a $5\frac{1}{4}$" bay. Developments such as read-write optical drives also demand the use of a spare $5\frac{1}{4}$" bay, and there may be many other add-ons that are under development that will require these fittings. Even the normal bog-standard arrangement of one hard drive and one floppy drive takes up two bays, so if you want to expand with a second hard drive or another floppy, a CD-ROM, a tape backup system (a streamer) and so on you will soon find the need to use space that was never intended for drive bays. If your case is of the miniature type you will have a buy a larger case and start again.

The older type of large casing looks very empty when the lid is raised, because modern motherboards are so much smaller than their predecessors. This makes more room for attaching other accessories, and it is not an advantage, as noted above, to opt for a small case unless you are very short of desk space. The conventional setup of a PC machine is to use the case to support the monitor, with the keyboard in front of both, but the cables that are supplied will usually allow for other configurations – for example I have the main cases for two machines on a shelf under the desk, with only the monitors and the keyboards on top. This confers the advantage that I can look downwards at the monitors without needing to jack up the chair to an unreasonable height.

The back of the casing should allow for at least as many slot openings as the motherboard uses, or, if possible, more. The number of the slots on the motherboard governs the number of add-on boards that can be used, and though you can easily estimate how many you need now, it is less easy to guess how many more you might want in a few months time. The number of slot positions in the casing determines how many boards can be used along with connectors that protrude from the back of the casing, and though several boards do not need connectors, it is always advisable to assume that each board will need to have a connector. On modern machines it is quite common for the IDE disk controller board to have several connectors wired to it and taken to the slots in the casing at other points, so that you may have more connectors than boards.

The hardware that comes attached to the casing is also worth looking at. Older casings will have a Reset switch and a Turbo switch, modern versions will have neither. The Reset switch is almost always at the front of the casing, and most users would prefer it to be at the back, well out of the way of fingers to avoid being pressed accidentally. The Turbo switch is a relict from the past when it was assumed that you might want to run the clock of the microprocessor at a lower speed at times for some games programs. Since the Turbo On position of this switch provides the speed that you have paid for, this is the setting that should be used at all times. After all, if you want to run games you can buy games machines. Once again, with the Turbo switch at the front of the casing there is a danger that its setting will be changed unintentionally, and my preference is to make the software ignore the switch setting if this is possible. Some CMOS RAM settings allow for this, but if this facility is not provided in the BIOS you will have to put up with the Turbo switch or alter the wiring so as to make the fast setting permanent. Very few motherboards come with explanations of the Turbo (and other) connections, so that you may need to work out for yourself what connections are needed to make the turbo setting permanent.

There should also be LED indicators for power on and for Turbo On, and usually another to indicate when the hard disk is reading or writing. The leads that are connected to these switches and LEDs are fairly well standardized, but I have come across a Turbo switch and LED which worked on a 286 motherboard but not on a 386 motherboard. If you, as recommended, ensure that the motherboard is always operating at full speed, the Turbo On LED is redundant.

Older cases provide a keyboard lock switch to disable the keyboard, though this has been omitted in later designs. Some machines will refuse to boot in this locked position, others will enter their CMOS set-up routine, a few will appear to start normally but will ignore anything typed. The differences depend on the instructions built into the ROM. The keyboard lock switch uses a very simple key, and should not be depended on as a way of securing the machine because virtually all keys are interchangeable and by opening the case the leads can be shorted to allow normal use of the computer.

The small loudspeaker which is used to provide warning notes is built into the case, and provided with leads that should reach the correspond-ing connectors on the motherboard. The location of the loudspeaker is not fixed, and different manufacturers are likely to place this component in different positions, but usually near the front of the case. The loud-speaker is of portable-radio standard, and if you want to use multimedia software which includes sound output you will in any event need to add a sound card to the machine. Such a card can be connected to external speakers or to an amplifier system that will drive loudspeakers of better quality (see Chapter 11).

The case must supply an opening for the main ON/OFF switch, which is part of the power-supply box. This may be placed at the rear right-hand side of the casing, but on more modern designs the switch is 'floating', connected to a cable so that it can be placed independently of the power supply position and allowing you to use cases that have switch openings at the front of the casing. Some of the slots that are provided for the main switch are a tight fit, particularly for a switch that is part of the power supply unit, and you may have to enlarge the slot in the case, using a Swiss file, so as to allow the switch to operate freely. If you are driven to using a file, make certain that all metal filings are removed before you fit anything into the casing.

Finally, the floor of the casing has pillars that will support a motherboard. Very often there are more pillars than the motherboard needs, and there may be some mounting-holes on the motherboard that do not correspond to pillar positions. This is seldom a problem, because provided the motherboard is well supported it does not need to be bolted down in many places.

The motherboard

The motherboard (Figure 2.3) is the main board that contains the microprocessor, its support chips and the main system memory. Early motherboards were large because the memory, up to 640 Kbyte in these days, consisted of a large number (typically 18 or 36) of individual

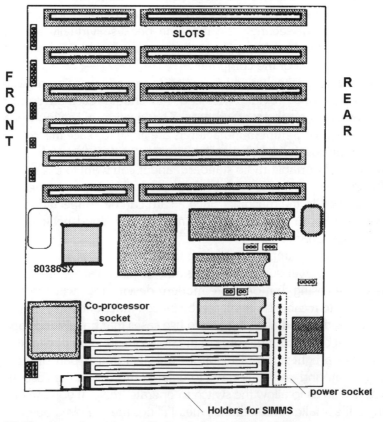

Figure 2.3 *The shape of a modern motherboard, showing the main features.*

Figure 2.4 *A Maplin motherboard, showing the slots on the right-hand side and the holders for SIMM memory on the right of this picture.*

memory chips, each with a comparatively small amount of memory. For some time in the late 80s and early 90s, motherboards were supplied with 1 Mbyte of memory in this chip format with provision for adding more memory by inserting SIMM or SIPP assemblies. Figure 2.4 shows a modern Maplin motherboard, reference ZG05.

SIMM memory, an acronym of Single In-line Memory Module, uses a strip of high-capacity memory chips, either three or nine chips, in a slim card unit, Figure 2.5. The SIMM makes contact with the motherboard

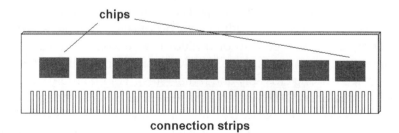

Figure 2.5 *A SIMM unit, which carries a set of memory chips and plugs into a socket on the motherboard. The illustration shows a nine-chip set, but the most recent SIMMs use only three chips.*

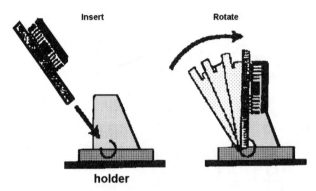

holder

Figure 2.6 *How a SIMM is secured in its holder. The SIMM is dropped into place at an angle, and then turned upright to lock it.*

through a set of thin metal strips which are inserted into a SIMM socket on the motherboard. The normal method of fitting nowadays is to slide the strip into its holder tilted at 45°, and then secure it by turning the unit until it is at right angles to the motherboard, Figure 2.6. This locks the SIMM in place and ensure that all the contacts are made securely. The earlier SIPP units used pins for connections and were plugged into a long socket, but there was no locking device to ensure that the unit did not vibrate out of place. By 1993, it was becoming quite difficult to find suppliers for SIPP memory, and where they were available they were more expensive than SIMMs.

Note that SIMMs exist in two main forms, 30-pin and 72-pin, with the 30-pin type used on older (mainly 80386) machines. The 72-pin SIMM is standard for 80486 and Pentium machines, and a few also use EDO memory, which can be mixed with 72-pin SIMMs. The motherboard documentation will show which type is used (and you can always count the pin socket positions). In addition, the 30-pin type of SIMM was at one time supplied as a three-chip or a nine-chip version. There are subtle differences between these types, and you should not mix the varieties (think of it as mixing cross-ply and radial tyres). Your memory should be entirely three-chip SIMMs or entirely nine-chip SIMMs, though with a modern BIOS you can get away with not mixing SIMMs in one bank (so that you could have one bank of three-chip and one bank of nine-chip). The safest course is to use only the more modern three-chip variety unless you are topping-up the memory of a machine which has already some nine-chip SIMMs fitted.

The more recent motherboards use no memory chips on the main board, and rely entirely on added SIMMs. This allow the boards to be sold in zero memory form (appearing as 0K in the advertisements, something that can puzzle buyers), with the memory added in SIMM form as needed. At the time of writing, the 1 Mbyte and 4 Mbyte SIMMs were normally catered for on 80386 motherboards, but the memory usually has to be installed using one type or the other, with no provision for mixing types. In addition, 30-pin SIMMs normally need to be installed in pairs, one pair forming a bank of memory. In a typical 80386SX motherboard, for example, this allows you to use 1 Mbyte SIMMs for 2 or 4 Mbyte memory, or 4 Mbyte SIMMs for 8 Mbyte or 16 Mbyte, with no provision for any other amounts of memory. 16 Mbyte is the memory limit for the 80386SX chip, though the 80386DX and the various 80486 chips can handle colossal amounts of memory.

For a modern motherboard that uses 72-pin SIMMs, you can mix all sizes, so that you could start with a 2 Mbyte SIMM, then add a 4 Mbyte later, and so on. There is, usually, no need to add 72-pin SIMMs in sets. On the most recent boards with EDO (Extended Data Output) memory, you can mix these memory units with 72-pin SIMMs. Note also that you can buy adapters that allow a set of 30-pin SIMMs to be placed into one 72-pin holder.

Slots

The slots on the motherboard, illustrated in Figures 2.3 and 2.4, are the connectors that hold and make connection to expansion boards. This follows the system used on the original PC machine and featured even earlier on the Apple computers of the late 70s. On a machine bought in 'bare-bones' trim, only two of these slots will be occupied, one with the disk controller card which will allow the use of up to two hard drives and four floppy drives. This card nowadays usually contains also a parallel port (for the printer) and a serial port (for a modem). The other essential card will be the video graphics card.

Slots are supplied long or short. The short slot uses 36 connecting pins and is intended for use with only the older type of expansion cards compatible with the XT computers. These short slots are either omitted on most modern motherboards, or the number is limited to one or two.

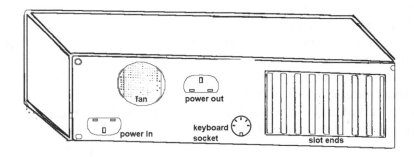

Figure 2.7 *The slot openings at the back of the casing. These allow connectors on cards to protrude from the casing when a blanking strip is removed.*

The long slot has a 62-pin connector and is arranged so that a short-slot card or a long-slot card can be plugged into it. It provides for 16-bit data signals and is more suitable for modern machines and plug-in cards. All of the motherboard slots are in positions that correspond to openings in the casing at the rear, Figure 2.7, which are normally covered.

All of these expansion slots operate under a handicap that has been inherited from the original XT machine. Whatever the clock rate used by the processor on the motherboard, the clock rate on the expansion slots is much slower, as low as 4.16 MHz on XT machines, though 8 MHz is used on some AT expansion slots. At one time this could be justified on the grounds that expansion cards carried out actions that did not need a higher clock rate, and certainly this is true of such actions as parallel or serial ports, older hard disk drives, floppy drives and the older forms of graphics cards. The slow clock rate on the expansion slots is now a disadvantage for fast graphics cards, and can also form a bottleneck for the most recent hard disk controllers.

This is where local bus slots come in. A local bus slot uses a much higher clock rate than the other slots, and can operate suitable cards at a more advantageous speed. The problem is that these slots are not entirely standardized, so that you cannot be quite certain that a given graphics card will operate in the local-bus slot on your motherboard. There are two main types of local bus slots, called VLB and PCI respectively. The VLB type is used mainly in the 80486 type of machine, and the PCI type for Pentium machines. If you are likely to upgrade from a 486 to a Pentium by changing the motherboard at some future date, you should use a motherboard with a PCI bus and buy boards (such as hard drive interface and video graphics board) that fit PCI. This will avoid the need to change these boards when you upgrade. Local bus slots are not normally found on 80386 motherboards so that this option exists only if you are upgrading or building a 486 machine.

The usual slot provision on recent motherboards is six full-length (16-bit) slots, with an additional two local-bus slots on the 80486 motherboards, or possibly a couple of short slots on boards intended for the 80386. Though you may initially use only two slots, one for the graphics card and the other for a combined disk controller and ports card, using a motherboard with too small a number of slots is a false economy. Motherboards that are offered for replacement purposes usually provide the standard count of six slots, but some machines come with as few as three.

The need for slots arises from the need for expansion. If you want to add a sound system, a CD ROM player, an optical floppy drive, a scanner or any of a host of desirable add-ons, each one is likely to require slot space for its controller. Once you have filled all of the slots your expansion capabilities are sharply brought to an end, because there is no easy way of providing for further expansion. Your only option then is to use expansion systems that act through the parallel port, using an adapter which still allows the printer to be used. This is not always a feasible scheme, because parallel ports on the older machines are strictly one-way, and only a two-way parallel port can be used for this type of expansion for external drives. In addition, devices which connect to the parallel port in this way are always much more expensive than those which simply slot in internally.

Power supply

The power supply unit (PSU) is the large box situated at the right-hand rear side of the computer, Figure 2.8, and prominently marked with notices that it must not be opened by unqualified personnel. Observe this warning, and if you must open the box first make certain that the machine is unplugged from the mains, that all cables have been disconnected, and that it has been switched off for at least ten minutes to allow capacitors to discharge.

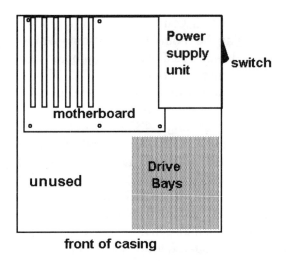

front of casing

Figure 2.8 *The position of the power supply unit (PSU) in the casing.*

The power supply box carries also the main switch and the fan, and it is connected (in the UK) by a standard three-pin rectangular socket of the type known as a Euroconnector. The plug end of this will be attached to a cable to which you can fit or have fitted a standard three-pin mains plug. The mains plug should carry a 3A fuse – do not on any account use a larger rating.

Power supplies are, almost universally, rated at 200 watts. This provides for a 5 volt supply at 25 amps, with +12V 5 amps and -12V and -5V at 1 amp each. This should be adequate for even the most heavily extended machine and for most users is much more than is needed, but the techniques that are used allow a 200W unit to be built at virtually the same price as the older 150W or 90W types. Some respected suppliers are still using 90 watt power supply units on 386 machines. If you have some experience of power supply units on other electronic equipment, note that the computer power units are of the switching type, and specialized knowledge is needed to deal with them if trouble arises. The one servicing exception is replacement of the main reservoir capacitor – this is usually the culprit if the unit fails after being switched off and on again several times in quick succession.

The fan that is built into the power supply unit is usually faster and noisier than is needed for cooling, and unless you have experience in working with electronics circuits there is nothing you can do about it because the fan is mounted inside the power supply unit. If you feel sufficiently experienced to open the power supply box, and prepared to accept the warnings, Papst can supply silent fans or a silencing unit for existing fans. If you have an urge to do it all yourself, you can wire a couple of thermistors in series with the fan's 12V supply, using ElectroMail (R-S Components) type TH-7 thermistors. For greater speed reduction try one single TH-3 thermistor in place of the two TH-7. There are no Maplin equivalents listed at the time of writing.

The power supply unit almost always includes an AC mains output, a 240V mains supply which is controlled by the computer's main switch. This is normally used either for the monitor or the printer (it can be used for both if you can wire them appropriately). The output comes from a fixed socket, and a matching plug is seldom supplied. Few electrical shops stock such plugs, and they are best obtained by mail order, using firms such as ElectroMail or Maplin. The ElectroMail reference number for a suitable straight plug is 489-251 and the Maplin code is HL16S. Business users with an RS Components account can order using the same code as for ElectroMail.

Keyboards

The original form of XT keyboard used 83 keys laid out with the function keys grouped at the left-hand side and the number keypad at the right. The number keys on the right also doubled as cursor keys with a number lock key used to avoid the need to hold down the Shift key when numbers were being entered from this keypad. This layout was widely criticized, and some of the points that were made were that the Esc key was too close to others, there was only a single Ctrl key at the left-hand side, and the combining of the cursor keys with the number keypad was inconvenient.

The later type of keyboard has 102 keys, laid out (Figure 2.9) in more logical groupings. The function keys are set across the top of the keyboard, with the Esc key isolated to the left. The cursor movement keys are now separated from the number keypad, as also are the Insert, Home, Page Up, Page Down, Delete and End keys. The Ctrl keys are duplicated so that they can be used with either hand, making it easier to press Ctrl with any other key.

What you regard as a good keyboard is very much of an individual preference. Most users like a keyboard to have a positive click action to

function keys

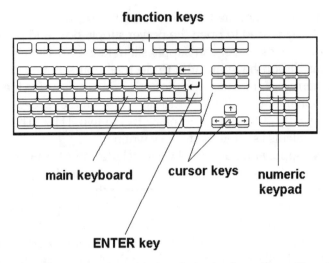

main keyboard

cursor keys

numeric keypad

ENTER key

Figure 2.9 *The modern 102-key form of keyboard, showing the positions of important sets of keys.*

the keys, but without excessive noise, and keyboards of the rubbery variety are universally disliked. Most anonymous keyboards are of a reasonable standard, and those branded with the name Cherry are highly regarded. Note that you cannot interchange the old XT keyboard with the later AT keyboard, though most AT keyboards include a switch that allows them to be used on XT machines. Early Amstrad machines used a non-standard keyboard connector. Figure 2.10 shows a Maplin keyboard whose action is of the Cherry type, with a good response from the keys – they feel right, as opposed to the dead feel of some keyboards. Since the keyboard is your main communication with the computer, a good keyboard is important.

Though keyboards of the later 102-key variety all look very similar, there are subtle differences. Some nameless clone machines come with US keyboards, which are easily detected because of the absence of the £ symbol (on the upper '3' key). Though a US keyboard can be used without difficulty, the absence of the £ sign can be irritating, though you can usually get the symbol by holding down the Alt key and typing the number 156 on the separate keypad, then releasing the Alt key. If you seldom need to use a £ sign, there is no problem.

A point that often causes confusion is the provision of an Alt key on the left of the spacebar and a key marked Alt Gr on the right-hand side.

Figure 2.10 *The Maplin ZG57 keyboard, a particularly good one for the constructor at an attractive price.*

The Alt Gr key is intended to be used on machines that make use of the German character set and a lot of PC software will allow you to make use of either the Alt or Alt Gr key interchangeably. This is not always true, however. If you are accustomed to switching between Windows programs by using the Alt and Tab keys together you will find that the Alt Gr key cannot be substituted for the Alt key. Microsoft Word for Windows also ignores the Alt Gr key (and also the right-hand Ctrl key) for many of its actions. The main use of this Alt Gr key is to provide the third character on keys that display more than two character symbols, and on UK keyboards the only character of this kind is the split bar | which is usually shown also on the key next to the letter Z. One curious effect is that the solid bar and the split bar are usually shown the other way around on the screen.

Keyboards should be kept covered when not in use, because dust can gather at an alarming rate. This can cause keys to jam, and when this happens you will see an error message appear when you try to boot up the machine. The message is 'Keyboard error' followed by a code number which shows which key is causing the error; for example 0E indicates that the Alt key is jammed down.

Another factor to consider is how keyboards age. Some keyboards do not change at all in the course of their life, others alter quite noticeably. One of my keyboards (a US layout) started life with a pleasant click action, but has deteriorated so that keys now stick closed or jam open at times.

The other keyboard has remained good, and the old Amstrad keyboard (pensioned off after four years and a million words typed) remains as good as new.

Never be tempted to spray WD40™ or any other silicone lubricant onto a keyboard whose keys have started to become sticky. Using a spray virtually guarantees that some of the liquid, which is one of the best insulators known, will get into contacts, ensuring that the keyboard will never work again. You might be able to release one sticky key by careful lubrication using a drop of silicone oil on a piece of wire, but don't depend on it.

Motherboard preparation

Before you can consider starting assembly, the motherboard needs to be inspected carefully, and you also need to read the manual or other documents that accompany it. If there is no form of documentation, contact the suppliers of the motherboard because you cannot assume that you will be able to find the correct connections or to make the correct settings by instinct.

Jumpers, Figure 2.11, are used to make contacts on the motherboard to switch actions in or out, or to allow for options. Each jumper unit normally consists of a row of three small pins with a bridging clip, the jumper itself, which can be placed over two pins to provide two settings (sometimes three settings if the design provides for the jumper to be removed altogether). Jumper settings should be correct if you have

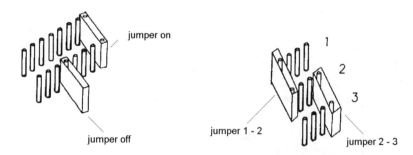

Figure 2.11 *Two types of jumper strips as used on motherboards and on cards for setting options.*

bought a bare-bones system with the motherboard already installed in its case, and very often there is little chance of altering jumpers once the machine is fully assembled.

If you are installing a new motherboard or replacing a motherboard, however, you will need to check the jumper settings very carefully before you place the new motherboard into the case. The small (and usually anonymous) manual or leaflet that comes with the motherboard will list the jumper settings, and these are usually preset correctly. If they are not, it is not always clear what settings you ought to use, and you may need to enquire from the supplier of the board. Another problem is that manuals usually show the pins numbered, but this numbering is not necessarily printed on the motherboard or, if it is printed, it is obscured by chips or other resident obstacles. The description that follows is of jumpers on a recent 80386SX board. This is fairly typical of modern practice, and most boards that you are likely to come across will provide for a similar list of jumpers.

The most important are the clock speed jumpers (usually two) which need to be set normally for the higher clock speed, typically 25 MHz or 33 MHz. These will normally be preset for the higher speed, but should be checked. Another jumper will determine whether the co-processor (if fitted) will run synchronously or asynchronously with the main system clock. If you fit a co-processor (see Chapter 7), you should be able to find from the supplier of the co-processor chip whether synchronous or asynchronous operation is required. This jumper is often located in a position that is very difficult to reach when memory and disk drives have been installed.

Another jumper needs to be set for the type of CMOS RAM chip. The two common types for 80386SX boards are the US1287 or the 146818, and since the manufacturer of the motherboard will already have installed the CMOS RAM chip, the jumper should have been correctly set (sometimes the default is to have no jumper and only two pins in this position).

Some motherboards use a jumper to set conditions either for a monochrome or a colour monitor. If you use a VGA graphics board it is better to set for a colour display even if you intend to use a monochrome VGA monitor. The monochrome setting is intended for the Hercules mono display system. If a parity-check jumper is provided, you can normally accept the default setting. Memory parity checking is a method of detecting memory errors, and it was considered important in the early days when memory was considered unreliable. On modern systems parity checking of memory is a waste of time, and the jumper can be set to disable it. This option is often available in the CMOS RAM set-up also. Parity checking *must* be disabled if the machine uses memory in 8-bit, as opposed to 9-bit, sets.

Some boards provide a jumper for address delay, with no explanations. You can either accept the default setting or disable address delay – very few machines are likely to need this setting to be switched on. Older boards are often provided with a much larger number of jumpers, and one common action that is provided is to enable or disable the on-board battery that powers the CMOS RAM while the machine is switched off. Older boards often included a connector which could be used along with an external battery, a useful expedient if the built-in device failed. This facility has been dropped by many manufacturers on the grounds that it was never used, and that motherboards are generally replaced long before the end of the battery life. Older boards also often used jumpers to enable or disable the floppy disk and the hard disk controllers. These also have now been dropped because they were never used, and in any case such jumpers belong on the disk controller board.

Take your time, enquire if necessary, and do not install the mother-board into the case until you are totally satisfied that the jumper settings are correct. A familiar problem is that the documentation may tell you that the setting you want is to jumper pins 1 and 2, but there is no pin numbering on the motherboard. If you come across this problem, you will often find that you can deduce pin numbers by looking at other settings which you are fairly sure have been correctly preset. You may find, for example, that pin 1 is the pin closest to the front of the motherboard.

Following the setting of jumpers you will need to install memory, and on all modern motherboards this is done using SIMMs. SIMM means Single In-line Memory Module, and is like a miniature expansion card containing a set of memory chips, as shown in Figure 2.5. Connections are made to the SIMM just as they are to expansion cards, using an edge-connector, a set of tiny metal tongues on the card which engage in springs on the holder. The older SIPP unit differs from the SIMM only in using a line of pins rather than the edge-connector; the name means Single In-line Pin Podule. The SIMM units are more common nowadays.

These SIMM units are inserted into their holders in the motherboard with the metal tongues on the SIMM pressing against the metal contacts of the holder. The standard SIMM units are of 1 Mbyte x 8, 1 Mbyte x 9, 4 Mbyte x 8 and 4 Mbyte x 9, allowing for expansion by 1 Mbyte or

4 Mbyte per unit. Some manufacturers provide 2 Mbyte x 8, 2 Mbyte x 9 and even 16 Mbyte x 8 and 16 Mbyte x 9 SIMMs. You will have to check with the manual for your machine to find which size of SIMMs it can use – the usual selection is 1 Mbyte x 9 or 4 Mbyte x 9.

This does not mean that you can expand a 1 Mbyte computer to 2 Mbyte by adding a 1 Mbyte SIMM. The usual modern arrangement, using banks of memory, is that SIMM units must be installed in twos, so that you can expand by 2 Mbyte at a time using the 1 Mbyte SIMMs and by 8 Mbyte at a time using the 4 Mbyte SIMMs. Check with the manual for your computer to find what arrangement is needed. Some machines may even require SIMMs to be installed in sets of four. Computers vary considerably in this respect, and you need to find from the manual what memory can be added and what changes to switches or jumpers may be needed. Some modern designs allow memory to be extended in 2 Mbyte units with no change to jumper settings. Make sure that the banks of memory (pairs of SIMMs) are installed in order – fill up Bank 1 before filling Bank 2 for example.

The later types of 80386 motherboards provide only for SIMM use in pairs, so that the minimum memory size is 2 Mbyte. The maximum is usually 16 Mbyte for 386SX machines, more for 386DX and upwards. Early types of SIMM sockets were built rather like expansion card slots, so that the SIMM unit was pushed straight into the socket. The more modern SIMM holder is more elaborate. The SIMM is slotted in at an angle of about 45° and then straightened up, when two plastic clips hold it in place. This locking arrangement is very much more secure than the older type, and takes the same SIMM units. You must be certain which type of holder your computer uses – machines made up to the first half of 1991 were still using the straight-push in type of SIMM holder, though some manufacturers had switched over. By 1992, the slide-in and twist type, noted in Figure 2.6, was predominant. It is important not to use the wrong method of insertion as this could damage the SIMM units and possibly the holders also. In addition, you should not mix three-chip and nine-chip SIMMs, certainly not in the same bank, because the chips have slightly different speed characteristics.

The later 80486 and Pentium boards will use 72-pin SIMMs, and these are much easier to work with, in the sense that they can be used in single units, though very fast machines required SIMMs installed in pairs.

Motherboard assembly

The motherboard must now be mounted into the casing, but don't rush into this task. Very often when a motherboard is in place, part of it lies under the power supply unit, and because of this, the PSU (Figure 2.12) is often supplied separately, not connected in. Don't connect in the PSU at this stage, and if the casing has come with its PSU fastened into place, check with the motherboard locating point to see if any of the mother-board will be covered by the PSU. If it is, as is normal, you must remove the PSU by unscrewing the three small bolts at the rear of the case and the three underneath. If your PSU uses a different number of bolts, make a note of this. A miniature socket set is useful for these bolts.

With the PSU laid temporarily out of the way, you can now concentrate on the motherboard mountings. Metal cases for the PC have their locating fasteners located in standardized positions, and motherboards are pro-vided with matching location holes, so that it is very unusual to find that there are any problems in fitting a new motherboard. Do not expect, however, that a new motherboard will have as many mounting holes as there are fasteners on the casing, or that all of the mounting holes will be in the same places. Remember, though, that a motherboard should *never* be drilled because the connecting tracks on the surface are not necessarily the only tracks that exist; most boards are laminated with tracks between

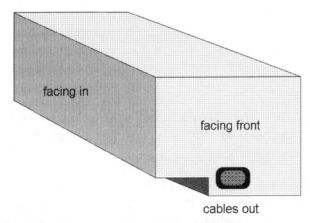

facing in

facing front

cables out

Figure 2.12 *The shape of the PSU box allows part of the motherboard to lie under the step in the base of the box.*

layers. Drilling through any of these tracks would be a very expensive mistake.

The fitting methods vary, but the most popular systems use either a brass pillar at each fixing position or a plastic clip, often some of each. The brass connectors are screwed into threaded holes in the case and the motherboard is bolted in turn to the pillars; the plastic clips that fit into slots in the case are pushed into the holes in the motherboard and then slotted in place. There should be at least one brass pillar fixing used to earth the motherboard electrically to the casing. Quite often, only two screwed fittings are used, with the rest being either clips or simply resting points. The motherboard should be supported under the slots, because this is where pressure is exerted on it when cards are plugged in. If there are no supporting pillars in this region you may be able to get hold of polypropylene pillars of the correct size and glue them to the floor of the casing – do not under any circumstances glue anything to the mother- board itself.

When you have the motherboard in place, check everything again. It is remarkably easy to plug in jumpers with only one pin making contact, for example, and when you come to make other plug and socket connections this is also a hazard to look out for. If the paperwork that came with the motherboard did not have a sketch of the motherboard, this is the time to make one that shows where the board is mounted and where the jumpers are. Remember that it is often very difficult to alter jumpers once a motherboard has been fitted in place, particularly if the jumpers are underneath the power supply box.

You now need to install the PSU box. If you removed it earlier you will know where everything fits, but if the PSU was not fastened into the case you will need to spend more time on this task. The first requirement is to check that you have all the mounting bolts – these are American UNF types which are not easy to replace, certainly not at your local ironmonger or DIY store. The second point is that the PSU has to be slid rather carefully into the casing, engaging the mains switch into the slot in the side of the casing and ensuring that the weight of the PSU does not rest on the motherboard. The thick and stiff set of cables from the PSU makes this task more difficult than you might expect. The shape of the PSU box allows part of the motherboard to lie underneath it without touching the components on the motherboard. Once the mains switch is located, it is easy to position the PSU so that all the screw holes line up, and the bolts can be put in place, finger-tight at first. If the PSU uses a flying mains

switch that will locate in the front panel, you can fit the PSU independently of the switch and clip in the switch later.

As you tighten the bolts, check that the mains switch can be operated
easily. Some casing slots are a tight fit for the switch, and if the PSU
mountings are a fraction out of line the switch will jam or be stiff. This
can usually be avoided by moving the PSU slightly on its mountings as
you tighten the bolts, but you may need to file the slot to get a perfect fit.
If this is needed, take the PSU out again, and file with the outside of the
slot pointing down, avoiding any filings landing on the motherboard. Tap
the casing afterwards to remove any lurking filings – just one filing
bridging tracks can cause puzzling symptoms and no-one can run a
diagnostic test and instantly exclaim: 'Yes, of course, you have a steel
filing bridging two tracks'.

rear

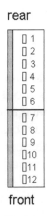

front

Figure 2.13 *The sockets on the motherboard for the power supply plugs. The two plugs are
often fastened together.*

Once the PSU has been installed the two main plugs must be inserted
into their sockets on the motherboard. These plugs are made so that they
fit together in a line, and they should plug in one way round only. When
you are replacing an old board, it is easy to mark the plugs so that you
can see which way round they go, but with nothing to guide you it is
considerably more difficult, particularly since the two plugs look almost
identical, Figure 2.13. If there is nothing else to guide you, the way that
the plugs are arranged on the power supply cable is usually a good clue
– if you have to bend, twist or rearrange the cables you are almost
certainly putting the plugs into the wrong positions.

Chapter 3

About disk drives

The original type of PC used a 5¼" diameter single-sided disk of 180 Kbyte capacity, but this was soon superseded by the familiar double-sided variety of 360 Kbyte capacity. Along with the AT machine in 1982 came the 5¼" floppy of 1.2 Mbyte, and the PS/2 range in 1987 started the trend to the use (pioneered by Apple) of 3½" disks, initially in the 720 Kbyte capacity. The standard form of disk now is the 1.44 Mbyte 3½" type. This latter type of disk is floppy only in the sense that the magnetic disc inside the casing is floppy; the casing itself is rigid plastic unlike the cardboard used in the old 5¼" designs.

The total adoption of 3½" drives is one of the few really noticeable external differences between the early PC/XT type of machine and the machines that started to appear at the end of the 80s. The same system of recording and replaying heads is used for hard disks as for the floppy type, though the spacing between the disk surface and the heads is usually much smaller. In addition, the larger-capacity hard disk units use multiple disks with two heads per disk. As Chapter 1 has pointed out, a PC must use both a hard drive and a floppy drive. This is a minimum, and there is nothing to stop you having more than one of each provided that you have suitable cables and driver cards.

The floppy drive

When you insert a 3½" floppy disk into a drive the hub spins briefly until the key in the drive shaft engages the slot in the disk, Figure 3.1. The act of inserting the disk also draws aside the shutter which normally protects the disk surface from air-borne contamination. When the drive is used, the drive motor starts to spin the disk at a speed of about 300 revolutions per minute.

The disk itself, Figure 3.2, is a circular flat piece of plastic which has been coated with magnetic material on each side. Through the slot that is revealed when the shutter is drawn aside, the heads of the disk drive can touch the surface of the disk, one head on each side. These heads

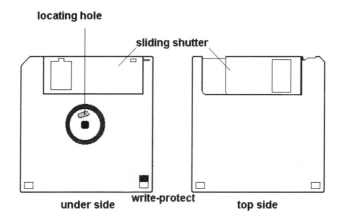

Figure 3.1 *The modern 3½" floppy disk is in a rigid container, with its surface covered for protection. The only difference between the low and the high density casing is the additional hole.*

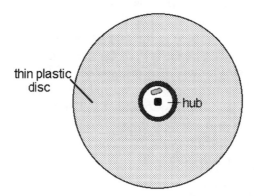

Figure 3.2 *The disk inside a 3½" casing is of thin plastic coated with magnetic material.*

are tiny electromagnets, and each head is used both for writing data and for reading data. When a head writes data, electrical signals through the coils of wire in the head cause changes of magnetism. These in turn magnetize the disk surface. When the head is used for reading, the changing magnetism of the disk as it turns causes electrical signals to be generated in the coils of wire.

This recording and replaying action is very similar to that of a cassette recorder, with one important difference. Cassette recorders were never designed to record digital signals from computers, but the disk head is (it magnetizes the tape fully in one direction or the opposite). The reliability of recording on a disk is therefore very much better than could ever be obtained from a cassette, which is why computers for serious use never feature the use of ordinary audio cassettes.

Unlike the head of a cassette recorder, which does not move once it is in contact with the tape, the heads of the disk drive move quite a lot. If the head is held steady, the spinning disk will allow a circular strip (sometimes incorrectly referred to as a 'cylinder', which ought to mean a collection of these strips on several disks) of the magnetic material to be affected by the head. By moving the head in and out, to and from the centre of the disk, the drive can make contact with different circular strips of the disk.

These strips are called 'tracks', Figure 3.3, and unlike the groove of a conventional record, these are circular, not spiral, and they are not grooves cut into the disk. The track is magnetic and invisible, just as the recording on a tape is invisible. What creates the tracks is the movement of the recording/replay head of the disk drive. A rather similar situation is the choice of twin-track or four-track on cassette tapes. The same tape can be recorded with two or four tracks depending on the heads that are used by the cassette recorder. There is nothing on the tape which guides the heads, or which indicates to you how many tracks exist.

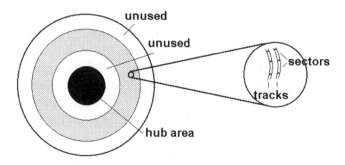

Figure 3.3 *The tracks on a disk are invisible, because they are magnetic patterns. The track area on a disk is usually on less than half of the total disk area.*

The number of tracks that you use therefore depends on your disk drives. The standard modern PC floppy-disk system uses 3½" disks with 80 tracks on each side. The tracks are packed to the tune of 135 per inch, so that 80 tracks occupy only about 0.6 inches of radius. These disks are known as high-density, abbreviated to HD. Once you have accepted the idea of invisible tracks, it's not quite so difficult to accept that each track can also be invisibly divided up. The reason for this is organization – the data is divided into 'blocks', or sectors, each of 512 bytes. A byte is the unit of computer data; it's the amount of memory that is needed for storing one typed character, for example. Each track of the disk is divided into a number of 'sectors', and each of these sectors can store 512 bytes of data. The 3½" disk will record 18 sectors of data on each track.

Looking at the arithmetic of this, using two sides each of 80 tracks and 18 sectors of 512 bytes per track gives a grand total of 2 x 80 x 18 x 512 bytes of storage, which is 1,474,560 bytes. If we take 1024 bytes per kilobyte and 1024 kilobytes per megabyte this comes to 1.406 Mbyte. These disks are usually advertised as being 1.44 Mbyte, a figure that is obtained by imagining that 1 megabyte is 1000 Kbyte rather than 1024 Kbyte. This is at least smaller than the factor that is used to boost the prices of US books on computing.

A disk called a *system* disk is used at the time when the computer is first switched on (booted). A floppy system disk contains the PC-DOS or MS-DOS tracks, thirteen tracks that are reserved for holding the hidden DOS files, leaving 67 tracks for your use. You would normally keep these MS-DOS tracks on the hard drive, however, so that on all but a few of your 3½" floppy disks only 33 sectors are reserved for the main 'directory' entries, leaving the rest free. This corresponds to a total of 1,457,664 bytes free on such a disk if you format a disk without copying over the MS-DOS files. This type of disk is a data-only disk. You would normally use such disks to contain word-processor text, data from spreadsheets or database or other programs, and programs which you might use along with others rather than in their own right. You would keep a couple of disks also which contain the MS-DOS hidden files and a few other files such as CONFIG.SYS and AUTOEXEC.BAT (see later) to use as backup system disks in the event of problems with the hard drive.

The next thing that we have to consider is how the sectors are marked out. Once again, this is not a visible marking, but a magnetic one. The system is called 'soft-sectoring'. Older 5¼" disks used a small hole, the sector hole, to indicate the position of the first sector on the disk. This sector hole is detected by the combination of an LED light source and a photo-transistor. The 3½" disk has a shaped drive-plate on one side, as illustrated in Figure 3.1 earlier. The drive hub has a pair of shaped pins that engage in the slots of the drive-plate, and since there is only one position in which the disk can be gripped, there is no need for a sectoring-hole in the disk and the first sector position is at a fixed point relative to the drive shaft. This must be a very rare example of a mechanical system replacing an electronic system. The software action called *formatting* will then place magnetic markings on the disk, taking this first sector point as a starting position.

When you load a program from a disk, or save data on a disk, you don't have to worry about the tracks and sectors. You don't, for example, have to specify at which sector and on which track the recording must start, and what to do if there is already something recorded on some of the sectors. All of this is the main action of the Disk Operating System (DOS), a sort of good-housekeeping system for disks, in conjunction with the disk controller circuits. The action of the DOS (see Chapter 8) is to ensure that a disk is of the correct format, to keep a record of what sectors are used for what files, and to allocate sectors as needed when data is recorded.

What you do need to worry about is the type of drive and disk that has to be used. In the development of PCs, advances in disk technology have lead to four main standards for floppy disks (the 3½" disk is still classed as floppy, and perhaps 'removable' would now be a better term). The older 5¼" disk exists as 40-track double-sided double-density 360 Kbyte, and as 80-track double-sided high-density 1.2 Mbyte versions – the capacity figures refer to a formatted data disk, and the higher-density disk format (meaning more bytes per sector) was introduced along with the PC/AT machine in 1982. The 3½" disk drives exist in 720 Kbyte and 1.44 Mbyte forms, with the lower capacity drives used mainly on older non-IBM machines. The bulk of the modern machines make use of the 1.44 Mbyte disk drive as the floppy drive, and rely on the hard disk for the main storage capacity. A few machines have appeared with a 2.8 Mbyte 3½" drive which can read both 1.44 Mbyte and 720 Kbyte disks. Unless you already have software on 5¼" disks you can regard this size as obsolete.

Problems used to arise in regard to distribution of software. Most software was once distributed on low-density 5¼" 360 Kbyte disks, with the option of 720 Kbyte 3½" disks, but later it became more common to distribute some programs, which were intended to run on only the larger and faster machines, on 5¼" 1.2 Mbyte disks. Nowadays, all software is distributed on 3½" disks, usually the 1.44 Mbyte type, and manufacturers will supply 5¼" disks only on request. If you are likely to want to make use of 5¼" disks, you can install a 5¼" high-density drive (which can read the old 360 Kbyte disks as well as the later 1.2 Mbyte disks).

CD-ROM is now a favoured way of distributing software, and even if CD-ROM were not used for multimedia it would be worth installing it for this purpose. The advantages include the reduction in weight – a CD is much more compact than 32 floppies – and avoidance of fraud, because the files on CD-ROM can be much larger than can be fitted on a floppy and therefore difficult to copy. If you buy software in this way you need not worry about making backups because corruption of a CD-ROM is much less likely than corruption of floppies.

Hard drives

The floppy type of disk has several limitations. The main limitation is that the disk spends most of its life out of the drive, subject to the dust and smoke in the room where the disk is housed. The recording and replaying heads are in contact with the disk surface, so that the fragile floppy disk cannot be spun at a very high speed because of the friction of the heads and of the sleeve which protects it.

The hard drive, originally called a Winchester disk, is a way of obtaining a much larger amount of information packed into the normal size of a disk. The older name of Winchester was used because this was the IBM name for a project that called for the development of a hard disk in a sealed unit in the early 1970s. Like other types of hard disks, Winchesters keep the disks themselves and the magnetic read/write heads sealed inside a container that ensures a dust-free environment for the disks. Unlike some other types of hard drive units, however, the Winchester is permanently sealed – the disks are not removable, and the sealing will be disturbed only if repairs are needed – and that will require an air-conditioned workshop. This sealing into a clean dust-free space allows the gap between each head and its disk to be made much smaller than could be contemplated for a floppy disk, smaller than a grain of dust or a particle of smoke.

The same system of using recording and replaying heads is used for hard disks as for the floppy type, though the heads are never in contact with the disk surface; they float on a thin film of air. In addition, hard disk units use multiple disks, made of coated aluminium, with two heads per disk, using anything from 2 to 8 disks (or platters) in a drive. The very small gap between the head and the platter allows a much higher packing of information on to the platter so that the most obvious effect of using a hard drive is the much greater number of bytes that can be stored. Figure 3.4 shows a typical arrangement of platters and heads. All of the heads are moved in step, so that at any particular time each head will be positioned over the same track on each side of each platter, and this set of tracks is called a *cylinder*.

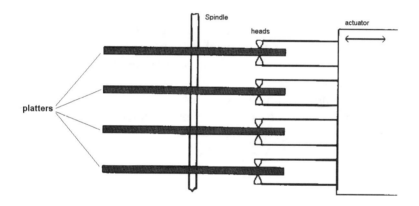

Figure 3.4 *Platters and heads on a typical hard drive. All of the heads move in and out together.*

Unlike the comparatively slow-spinning floppy disk, the hard drive spins at a very high speed, around 3600 revolutions per minute. This means that the rate at which data can be written to the drive or read from it is much greater, some twelve times as fast as a floppy disk. This speed of recording or recovery of data has been helped by the progress in reducing the size of these hard drives. The first units used 14" drives, and by the late 1970s 8" units were in full production. The modern 5¼" hard drive started to appear by the early 1980s, and the 3½" size in the mid-eighties, but it is only recently that prices have dropped to a level that allow really widespread use of hard drives on small computers.

The advantages of the smaller size have been lower costs, and faster access to data, because the head needs to move over a much shorter path on a 3½" platter than on the old 8" platter. Even smaller hard drives are now available, and form the basis of the hard drive on a card that allows a floppy drive XT type of machine to be upgraded to hard drive use with the minimum of installation and without removing an existing drive. The hard card units that are now available use twin disks with four heads, and about 615 tracks on each surface. The sectors are also recorded to a much greater density of data, with 2 Kbyte per sector rather than 512 bytes (0.5 Kbyte) as on the ordinary floppy so that fewer tracks are needed for a long program, and some programs will fit within a single sector. The sector size also means, of course, that a ten-byte piece of data will occupy a single sector with nothing else recorded there, unless you make use of a program which optimizes the use of disk space.

An important difference between floppy drives and hard drives is that the recording method for a floppy drive is standardized so that a floppy recorded on one machine can be used on another machine. This is not true of hard drives because a hard drive normally spends all of its life in one machine and is not moved around (though there are some removable hard drive units which use a standardized system). The hard drive and its controller circuits can be taken as one unit which *is* interchangeable.

Recording methods

Basically, the writing action of the disk head consists of magnetizing the surface of the spinning disk with a set of signals that correspond to the data that is being recorded, and reading the disk is done by the reverse action, allowing the varying magnetization of the disk track to affect the coil of wire in the reading head, providing the electrical signals. There are, however, many possible ways of coding electrical signals into magnetization of a disk track, and you will find references in older books to systems called FM and MFM. As far as the IBM PC type of computer is concerned, the system that is used for floppy drives is MFM (Modified Frequency Modulation), since this allows signals to be recorded at twice the density (signals per inch of track) of FM recording on the same disk. The system is also referred to as IBM System 34, and was used for early hard drive units also. Modern hard drives are free to use a variety of coding methods, because the hard drive unit contains its own coding/ decoding circuits. On older machines, it was essential to know what system was being used so as to ensure that the correct drive card was fitted.

Another difference between drives is the choice of stepper motors or voice-coil head drives. As the name suggests, a stepper motor moves the heads in or out by a fixed amount each time the motor is actuated, so that the setting of the stepper motor determines the distance between tracks. The voice-coil mechanism is similar in principle to the mechanism of a loudspeaker (hence the name) and allows track spacing to be set by electronic rather than mechanical methods, so that software can be used to change the number of tracks and their spacing. This latter type of drive is quieter and more versatile, but is more expensive to construct because of the elaborate control system that is needed to ensure that the heads are always perfectly aligned on the tracks. All modern hard-drive units should use a voice-coil mechanism.

Simply having a disk drive in place is not sufficient, however, because hardware and software is needed to make the disk operate. On the old PC machines, the disk driving hardware was on a plug-in card separate from the disk drives. On modern machines, the driving circuits for the hard drive are on the drive itself – hence the name of Integrated Drive Electronics (IDE) for this type of drive. Computers using this type of hard drive, more than 90% of all PC machines, have an IDE interface card which usually includes floppy disk drivers for up to four floppy drives, along with interfacing for up to two hard drives and ports for printer and serial connectors. A later version is called EIDE.

Using IDE, there is a data cable which connects from the IDE board to the hard drive (or to each of two hard drives). A separate cable from the power supply unit is used to provide power for the motors of both the hard and floppy drives. If you are likely to use more than two hard drives, or if you need to use a large number of drives of various types, you may need to consider the use of a system called SCSI (Small Computer Signals Interface), but if you are building a computer for yourself it is unlikely that the use of multiple hard drive units is a priority, and we'll leave this type of complication aside. The main reason for connecting a SCSI controller to a modern PC is that you are fitting CD-ROM or some other add-on that must have a SCSI interface.

Older machines are likely to contain disk controller cards which have to be matched to the type of disk. This used to be a source of considerable confusion, but the almost universal use of IDE has eliminated the problem. If you are re-arranging an old machine rather than building a new one, the important point is that the disk controller and disk drive must match up, so that if your disk drive requires, for example, an ESDI or RLL controller, then it must have one, and if you are fitting a new disk drive or a new controller you need to be aware of this. It makes more sense, however, to replace both with a modern IDE drive and interface board, not least because old units are not likely to have much life in them, and because the modern units are quieter and faster.

The most recent motherboards often contain the hard drive controller as a built-in part of the motherboard, and the manual for the mother-board will make this clear. If this interface is supplied, you will need to connect the data cable for the hard drive from a point on the motherboard rather than from a point on a separate driver card. You may still need a driver card for floppy drives, depending on the motherboard design.

The disk drive and the disk controller have to be able to exchange signals which determine what can be done. For a floppy drive, for example, there will be a Drive Ready signal which indicates that a disk is present and that the door of the drive is fastened. There will also be a Write Protect signal which will be sent from the drive if the write-protec-tion shutter of a disk is in its protected position, and a Dual-sided signal to show that the disk is a two-sided one. This latter signal is a hang-over from the past, when early models of IBM PC could use single-sided disks. For any type of disk drive, there have to be signals from the drive to indicate a Write fault, the position of Track or Cylinder zero and so on. The controller also needs to send signals to the disk drive, consisting of drive selection (A, B or D, for example), track/cylinder selection, sector selection, and head loading (for a floppy drive, meaning bringing the head against the disk surface). The controller also has to decode the data that is being read and encode the data that is to be written, and carry out error checks. The digital signals used in the computer are not used for the disk drive heads, nor are the signals from the disk directly usable by the computer. MS-DOS 6.0 onwards provides for a software system called DriveSpace that compresses the size of files before saving and expands before loading, carrying out actions that some day will be built into the drive controller.

Error checking is carried out using a method called CRC, cyclic redundancy checking. Each sector of data consists of numbers, and from these numbers a check number can be calculated. A simple scheme would consist of simply adding up all the numbers in a sector and using the total; but the schemes that are used are not so simple. This number is recorded with the data, and when the data is recovered the check number is also recovered and a repetition of the checking operation should produce an identical number. If this is impossible, the system must produce a suitable message on the screen.

The disk controller is under the control, in turn, of the DOS, so that the sequence of any disk action is a command to the DOS, which issues signals to the disk controller, which in turn issues signals to the disk drive. This will be followed by a transfer of data in one direction or the other. Whichever way the data flows, it normally uses memory buffers, so that data can be recorded or extracted in units equal to the sector size of the disk rather than in smaller units.

You may see references in books or in manuals to interleaving. The hard disk spins rapidly, around 3600 rpm, and it can happen that the old type of disk controller cannot transfer data between disk and computer fast enough to keep up with the disk speed. This problem is solved by a system called interleaving. For example, when a file is written, an interleaved system will write one sector, allow another to pass, then write on the next sector, so giving the system some time between successive writes, the time it takes for a head to pass over one sector. This time would typically be about 52 thousandths of a second, and such a system would be described as a 1:2 interleave. Slower systems might need 1:3 inter-leaving (write one, pass two) or 1:4 (write one, pass three) for satisfactory performance. When the sectors are read, the same system of interleaving is used in order to read the correct sectors. The use of interleaving obviously slows down the rate at which information can be transferred between the computer and the disk, and a 1:1 interleave is highly desirable. Once again, the controller and the disk drive need to be well matched. Modern machines all use a 1:1 interleaving, because the controller circuits are now on the disk drive, and can cope with the speed that is required.

Advantages of hard drives

Unless you have had previous experience of using a hard drive, it isn't exactly obvious what advantage other than sheer size of storage space they have. To start with, hard drives are essential for modern programs, which are too bloated to fit on a single floppy (some need six or more $3\frac{1}{2}$" disks, and a few even have to be distributed on CD-ROM). In earlier times, some popular programs, run from floppy disks, required you to change disks whenever you needed to change to some other function (such as from word entry to spell-checking or mail-merge on a word processor), and it was a considerable drag to have a program split up in this way. By transferring all of the files on to one hard drive, the entire set of files became available to you as soon as you switched your computer on. The files can also be accessed much more quickly, because there is no need to shuffle disks about, and the access time for the hard drive is so much faster than that of a floppy.

Even with the large size of modern programs, you can keep all of your business programs such as word processor, spreadsheet, database and ideas processor, on one hard drive, and switch from one to another in less time than it would take to insert a floppy disk. Even more useful, you can keep all of your MS-DOS system disk contents, and arrange for this to be loaded at the time of switch-on, and still leave room for data if you wanted to do so. For a lot of purposes, though, it's better to use the hard drive for programs, and the floppy disks for data, apart from data that is unchanging and which needs to be used frequently. With the larger hard drive, of course, there is ample space for data as well as for program files, but you will still probably prefer to keep data on floppies, if only from the point of view of having several backup copies. Since you should have the original program disks to act as program backup, using floppies to back up data can often make any more elaborate backup schemes redundant – the best form of backup is a simple method that you always use rather than a complicated one that you tend to shy away from.

You are not limited to one hard drive, incidentally, because you can still install a second hard drive into a floppy drive bay, provided you have a suitable data cable to feed two hard drives. The main hazard here is that the data cables are often so short that you need to consider the positioning of the two drives and the controller card very carefully.

Fitting a hard drive

We need to look at the installation of a hard drive first, because it is normal to keep the hard drive in the lowest of the drive bays, making it inaccessible once the floppy drives have been fitted. A built-in drive is the logical method of adding a hard drive to any machine which is of the standard PC type of construction, with a set of drive bays at the front of the casing. It is also the obvious system for slimline cases, provided that there is space, and for the tower type of unit if you can gain easy access.

The ordinary full-sized desktop case, particularly if it uses the conventional flip-top lid, is by far the easiest to work with, even if it does take up a lot of room on the desk (a bigger desk is usually cheaper than a smaller computer). The space that you have available in the case may dictate the physical size of hard disk you can use. In particular, a slimline casing may require you to look for a half-height or smaller style of drive, usually in the 3½" size of drive. It is most unlikely that any modern conventional casing will accommodate full-height drives, and you should avoid any bargain offers of such (usually old) units.

Do not assume that a drive will be provided with mounting brackets at exactly the same places as the drive-bay, though these positions are usually standard on PC clones. An adapter will almost certainly be needed, because most modern hard-drive units are of the 3½" size, and all but a few drive bays are 5¼" wide. The exception is the sideways type of drive bay which is often provided for a single 3½" hard drive, and this is a very convenient feature which can be used for a 3½" floppy drive if required. You should enquire when you order the drive what provisions are made for mounting it on the style of casing you are using. This is particularly important if you are upgrading an old machine because some manufacturers, notably Amstrad, use mounting systems which are very different from the IBM type of design.

The drive bay normally has slots at the sides to allow for to and fro adjustment of a drive, and two sets are usually provided at different heights in the bay. These should fit the adapter plate without any problems. Hard drives must be mounted to the adapter plate by way of small bolts fitting into their threaded mounting-pads. This is important because these pads act to cushion the drive against shock. In no circumstances should you consider drilling the casing of a hard drive in order to mount it in any other way. You should also handle a hard drive by its casing, not holding its weight on any other points. In particular, avoid handling the connector strips at the rear of the drive or any of the exposed electronic circuits.

The older 5¼" hard drives are provided with mounting pads at the side, Figure 3.5. This makes it easy to fasten them into the standard type of bay, which has slots cut in the sides for the mounting bolts. The 3½" and smaller drives use underside mountings as well as side mountings. This allows them to be screwed to the adapter frame (Figure 3.6), which has side pads that in turn can be fastened into a bay. If you have problems, Meccano brackets and strips can usually ensure that you get the drive unit firmly fastened. In a desperate situation, there is nothing wrong with

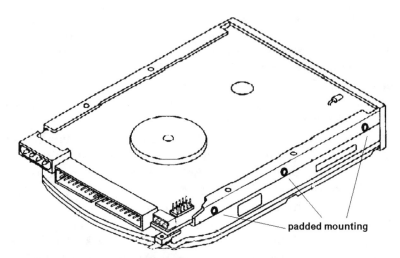

padded mounting

Figure 3.5 *Mounting pads at the side of a hard drive. Never be tempted to use any other mounting points, and certainly do not drill any mounting holes for yourself.*

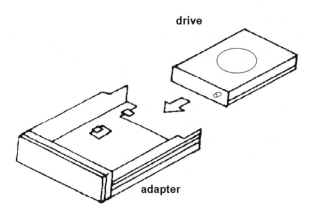

drive

adapter

Figure 3.6 *How a 3½" drive (hard or floppy) is fitted into an adapter case for mounting in a 5¼" bay.*

fastening the drive to a metal plate and sticking this to the casing with self-adhesive foam pads.

Jumpers and switches

The simplest possible installation of a hard drive is as the first hard drive in a machine that has only one or two floppy drives; or the replacement of an existing hard drive with an identical type. Complications arise only when a second (or further) hard drive is being installed, or when there are uncertainties about the compatibility of parts. The methods that are required vary according to the type of drive that is being fitted, and in this book we shall concentrate on the IDE type which is standard on modern PC machines.

An IDE hard drive can be installed as a first or a second drive. In normal circumstances, these will correspond to drive letters C or D respectively. The complication here is that these letters, known as logical drive letters, are assigned by the computer automatically, with both A and B reserved for use with the floppy drive(s), whether you have one or two floppy drives. The first hard disk will be assigned with the letter C only if this letter is not being used by anything else.

Older versions of the operating system, MS-DOS 3.3 or earlier, could not cope correctly with a hard drive of more than 32 Mbyte (now regarded as a hopelessly small-capacity hard drive). With these old versions of MS-DOS the machine had to use a drive larger than 32 Mbyte in sections called partitions, with a different drive letter for each. It was therefore possible for a single hard disk to be referred to as logical drives C and D or C, D and E. Modern machines use Version 5 or 6 of MS-DOS, and partitioning is not relevant.

Rather than talking about drive numbers or letters, though, it is preferable at this stage to talk about first and (possibly) second hard drives. Some manuals will refer to these as hard drives 0 and 1. When you install a single IDE drive on a machine, it should be configured as the first hard disk. This means that jumper settings have to be made to ensure that the hard disk signals are taken from the correct point in the computer, so that the operating system can make use of the disk. In technical terms, this is done by selecting the correct BIOS address number and the correct port address range on the controller board (see later). The settings are made by way of jumpers or DIP switches. Changes are needed on the controller only if a second drive *controller* (not just a second hard *drive*) is being used. For a first hard drive, these settings are almost always ready-made for you, and you need only check them. For a second hard drive, alterations will have to be made unless the suppliers have done them for you.

The main complication arises if you are fitting a second hard drive in a machine which has used a single hard drive. You will need to alter jumper settings so as to configure the second drive as a slave of a pair of drives. You will also need to take out the first hard drive (if it is already fitted) and configure this as the master drive of two. You will also need a data cable that has two hard drive connectors and which is long enough to reach drives that may be some distance apart. The documentation accompanying an IDE drive is often very sparse, no more than a sheet.

If you are installing a second IDE drive in an existing computer, use a drive from the same manufacturer as the first drive, and cables to match. This will help to avoid any problems of incompatibility. If this is not possible, enquire of the suppliers to check that the new drive you intend to fit will be compatible with the first type. This is normally not a problem with modern drive types, but some very old types are often sold at bargain prices.

The controller board

If you are adding a second drive, a controller board will already be fitted to the computer. A lot of 386 and 486 machines, however, are nowadays sold in bare-bones form, with a floppy drive but no hard drive. These machines always have a hard-drive controller board, though, and it is almost invariably an IDE type which can be used with one or two drives.

If you are using a 486 board or you intend to upgrade later, you should buy and fit a hard drive board of the EIDE type (extended IDE), because this will permit faster use and allows you to connect more than two hard drives. You may not want to use more than two hard drives, but the facilities of the EIDE board will be useful if you use it for connecting up the CD-ROM drive as well as a hard drive, leaving two more connections for another hard drive and for any other drives that may become available in the next few years. An EIDE board will use a VLB or PCI bus; there would be no point in using it on the old-style system bus slots.

Most IDE boards are supplied with one hard disk connector and a cable that connects to one drive only. If you want to use two hard drives, check with the supplier of the disk to obtain a new cable and to find what alterations are needed so as to run the second disk as a slave of the first. Remember that some modern motherboards include the EIDE circuits and connector.

A new controller board should be unpacked and checked, paying attention to any instructions that are enclosed. These may be very brief, particularly for an IDE board. If more than one drive is being used from the controller it may be necessary to make changes to the settings. Settings will certainly need to be changed if more than one controller board is being used in the same computer. The correct setting may call for no jumpers to be used, in which case a jumper is often supplied attached to one pin only and can be left that way. Similarly, any switch settings may all be to the zero side.

Installation

Before you start, check the drive package to make sure you have all of the mounting bolts, any adapter that is needed, cables (if not already on the computer) and instructions. Check that you have the necessary tools – a Philips screwdriver (possibly also a plain-head type) and a pair of tweezers are usually needed. The bolts are either 6-32 UNC x 0.31 (⁵⁄16") or metric M4 x 0.7-6H. The frame of the drive may be stamped with M for metric or S for UNC. If you need spare UNC bolts you will need to contact a specialist supplier, but the M4 metric types can be bought from electronics suppliers such as the well-known Maplin or RS Components.

At this stage, check that any jumpers or switches are correctly set. Once the drive is in place these will be impossible to reach. Use tweezers to manipulate these devices. It is not always obvious from the accompanying instructions what settings are needed, and though drives are often set ready for use in a standard 80386 type of machine you cannot rely on this. Jumpers will quite certainly need to be set if you intend to use more than one hard drive.

Unpack the drive carefully and read any accompanying manual thoroughly, particularly to check any prohibitions on drive fastening or mounting positions. No drive should ever be mounted with its front panel facing down, but most drives can be placed flat, or on either side. Check that any adapter plate fits into the mounting bay on the casing, and that all bolts and cable adapters (see later) are provided. The hard drive is usually placed as the lowest in a set of drives on a desktop casing. Check also that the drive data cable will reach from the interface (controller) board to the drive – you may need to put the IDE board in a different slot if the cable is short (as they usually are).

Fasten the 3½" drive to its adapter, using the small bolts that are provided to bolt into the mounting pads. Tighten these up evenly and not excessively. Handling the drive by its casing or by the adapter plate, place it into the mounting bay and check that the slots in the mounting bay match with the pad positions in the drive (5¼" drive) or in the adapter plate (3½" drive). Check that you can still place a floppy drive above the hard drive unit. This latter point is important, because floppy drives have an exposed flywheel on the underside, and the slightest contact against this flywheel will prevent the floppy drive motor from spinning. There should be no such problems if the 3½" drive is being mounted sideways in a bay specially provided for this purpose.

Place the bolts by hand and tighten evenly. Check as you tighten the bolts that the drive is positioned correctly. 5¼" drives usually have their front panels flush with the computer front, in line with the floppy drive(s). 3½" drives are usually recessed, with a plastic cover at the front of the case, or sometimes no break in the casing at all.

Installation is not a particularly skilled operation, though experience with a Meccano set as a child is helpful. Problems arise only if the mounting pads on the drive do not correspond with openings in the bay, or you have no adapter for a 3½" or 2.5" drive, or an unsuitable adapter, or you manage to lose a mounting-bolt. A mounting-bolt that falls inside the drive casing or the computer casing can usually be shaken out. Do not use a magnet to retrieve a bolt from a disk drive casing. Do not attempt to make use of other bolts, particularly longer bolts or bolts which need a lot of effort to tighten (because they are ruining the threads in the drive). It is better to mount a drive with only three bolts rather than to add one bolt of the wrong type.

IDE interface

The hard drive interface board is inserted as a plug-in card, using any available slot in the motherboard that will accept it. You may find that only one position is possible for this board because of the limited length of the data cables that have been supplied. The problems of length of cables can be more pressing if the IDE interface is built into the motherboard and therefore impossible to move. Figure 3.7 shows a typical IDE card which allows a generous cable length for the disk drives, and also contains ports and their connectors.

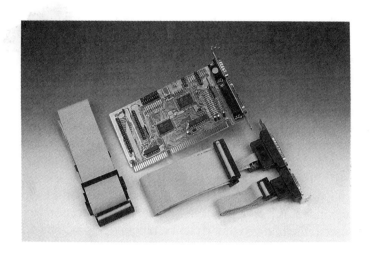

Figure 3.7 *A Maplin IDE card with provision for two hard drives, two floppies, two serial ports, a parallel port and a games port. The connectors for the ports are shown also.*

Make sure that all settings have been made on the controller board, and that its cables are fitted in place. Select a suitable slot for the controller board as close as possible to the drive(s). Remove the metal blanking strip (Figure 3.8) that is screwed to the end of the slot. Do not lose the blanking strip or the fixing screw.

hole for
retaining screw

notch to hold
lower end

Figure 3.8 *The metal blanking strip that normally covers a slot-hole in the casing is removed to allow the controller board to be put into place.*

Locate the controller board so that its connecting tongue is over the socket on the main board (motherboard) of the computer in the slot that you will use. Make sure that the tongue is lined up on the socket. The long 16-bit type is more difficult to work with. Press the controller board down, rocking it gently back and forth until it slips into the socket. The metal bracket on the back of the board should now line up with the retaining position that was used for the blanking strip.

Check again, and then use the fixing screw to retain the board. Do not over-tighten this screw, it serves only to ensure that the board does not vibrate out of place. Now connect up the cables to the drive(s). There are two sets of cables required for any hard drive, the power cable and the data cable set. The power cable is a simple four-strand type with a four-way connector (some drives use only three connections of the four). This connector is made so that it can be plugged in only one way round. The same power cable is used for floppy drives and for hard drives, and modern AT machines usually provide four or five plugs on the cable. The plug is a tight fit into the socket and usually locks into place. The socket for the power plug is obvious, Figure 3.9, but some disk drives need an adapter, which should be supplied.

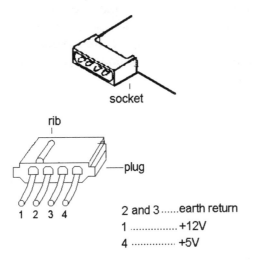

Figure 3.9 *The power socket and plug for either hard drive or floppy drive.*

The data cable that connects the controller board to the IDE drive is of the flat type. This plugs into the matching connector on the controller board at one end and into the drive at the other, with no complications. Look for one strand of the cable being marked, often with a black line, to indicate pin 1 connection. This makes it easier to locate the connector the correct way round. Do not assume that one particular way round (such as cable-entry down) will always be correct, or that a second hard drive will have its pin 1 position the same way round as your first hard drive. The power connection is as for any other system.

When you have the hard drive running satisfactorily (see later) it is desirable to mark the cable connectors so that you can replace them correctly in the event of having to remove the drives for servicing. Use Tippex or other white marker on the top side of each connector and write on the use (DATA1, DATA2, POWER1, POWER2 and so on). Mark also the pin 1 position on the cable and on the drive.

Checking out

When a hard disk, whatever the type, has been installed so that all the relevant steps described above have been carried out, you can check that the disk is mechanically capable of use. First, however, you need to make the machine ready for use. Plug the keyboard connector into its socket on the motherboard – this is usually a DIN-type socket located at the back of the machine close to the PSU. Insert the video graphics card that you intend to use, easing the card into its slot and screwing it into place. Plug the monitor data cable into the socket on the graphics card. Insert the monitor mains plug – if this is a Euroconnector it can be plugged into the socket on the PC main case, otherwise use a standard mains plug for the moment.

With all cables plugged into their correct places and the lid shut, switch on the power. If the monitor is separately powered make sure that it is plugged in and switched on. You should hear the high-pitched whine of the hard disk drive motor start and settle to its final speed.

If you hear a lot of disk activity and the machine boots (possibly with some error messages) then the hard disk is already formatted, and the formatting steps noted in Chapter 8 can be ignored. Congratulate yourself – you have avoided several tricky steps. This is as far as you can go for an unformatted hard disk, because you cannot use the drive until it has been formatted, but IDE drives which have been pre-formatted can often be put into service with little additional effort. If this is a second drive you have added, you can check that the machine is still booting up correctly from the first drive.

Problems, problems

At this stage, unless the machine has booted from the new drive, you do not really know whether you have any major problems, because all you can tell is whether the hard drive motor is running or not. If there is no sound from the drive, particularly when you are using a single hard drive, then the drive motor is not running. Check the cables if the drive has just been installed or replaced. This requires you to switch off, disconnect the mains lead, remove the monitor and open the lid.

It is most unusual to have this problem, because the power supply cable can be inserted only one way round. It is possible, however, that if an adapter has been used it is incorrectly wired or that a wire is broken. Check also for any signs of a break in the power cable, particularly at the connector.

Another cause of a hard disk not starting is a head jamming slightly against a platter. This should *never* occur on a new drive, particularly on any modern type with self-parking heads, but if you have bought or transferred an old (to the point of being ancient) card unit the problem can sometimes arise. The motor unit of a hard disk has a very low power output, particularly when it starts, so that only the slightest amount of friction is needed to make it stick. Even the metal strap over the drive shaft, which is used to prevent electrostatic charging, can exert enough force to make a motor reluctant to start.

The remedy, which sounds drastic and must never be used on a new drive, is to switch off, tap the hard drive gently and switch on, tapping again just at the instant of switch-on. This symptom can be a sign that the unit is failing, but very often the life can be extended very considerably by simply parking the disk heads (a unit as old as this will use a stepper motor, so that the heads are not parked automatically) each time the machine is to be switched off – you may never have the problem again.

Fitting the floppy drive

So much of the installation of a floppy drive follows the same pattern as fitting a hard drive that very little needs to be said here. Fitting 3½" drives into a 5¼" bay is done by way of adapter kits, and most machines will require at least one such conversion, because both the hard drive and the (only) floppy drive are likely to be 3½" types. The exception is when you install the floppy 3½" drive in a sideways bay. As before, take great care never to lose the fixing screws for these conversion holders and for drive bays, because they are types that are not easy to replace unless you have access to a computer shop with a good selection of hardware. Never assume that because a bay is provided this means that the cables supplied with the machine will be able to reach a drive added to that bay. Cables are often supplied that are so short as to restrict your layout seriously.

Check in particular that there is clearance between the underside of the floppy drive and the drive beneath it. This is not problem when the hard drive is fixed on its side in a bay intended for this purpose, but when you need to fit a hard drive and two floppy drives into a set of three bays you may encounter problems. In particular, if you are adding a 5¼" floppy drive and a 3½" floppy drive to a 3-drive bay that has a hard drive in its lowest bay, you may need to try the floppy drives both ways round before you find a position in which both will work acceptably. The problem, as noted earlier, is that the flywheel of a floppy drive is on its underside and can easily be fouled by any slight projection from the drive below it.

That apart, the main points to note are that the floppy drives have their jumpers correctly set (one will be drive 0 or A, the other 1 or B), and that the power cable is correctly used. Power cables nowadays are fitted with two types of plug, one for the older 5¼" drives and a smaller type for 3½" drives. The plugs are easy enough to insert, but it is not always easy to ensure that they are inserted correctly with all pins engaged. It is remarkably easy to insert a power plug with each of its pins against a piece of insulation rather than against the metal of a socket.

On the older versions of power units, all the power plugs are of the larger type, and an adapter (Figure 3.10) is needed to fit to 3½" units. This is straightforward, but if you do not have the adapter then you cannot proceed until you lay your hands on one. A good computer shop will often have some in stock. Remember to ensure that the connector to the 3½" drive is correctly inserted. The data connector should be plugged in the right way round, using the pin-1 marking on the data cable as a

power plug,
adapter

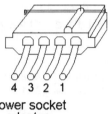

4 3 2 1

power socket
adapter

Figure 3.10 *The form of adapter needed for a 3½" power supply cable when the PSU cable contains only the larger plugs. Modern power cables are usually made with several sets of both plug types.*

guide. Do not assume that the plug goes in with the cable facing down – this can vary from one cable to another. Inserting the data plug the wrong way round has not caused any damage when I have tried it, but the disk system does not work.

Testing a floppy drive is easy enough. With the machine set up with monitor and keyboard (see above), place an MS-DOS boot disk (see Chapter 8) into the drive that is to be the A drive (usually a 3½" type). Switch on, and wait to see evidence of activity from the drive. The machine should boot up if all is well. If you have fitted two floppy drives and have booted up from one, try a disk in the other and check what happens when you request a listing of files by typing:

    DIR B:

and pressing the Enter key.

Section II

Improvers and modifiers

Chapter 4

Monitors, standards and graphics cards

Monitors

On the standard desktop type of computer, the image is produced on a video monitor, a phrase borrowed from television to mean a box which will produce a TV-style display on a screen of 12" diagonal or more. The details of monitors will be dealt with later in this chapter, but it is important to realize that a given monitor cannot be used with every type of computer graphics board because the types of signals have to be compatible. In particular, a monitor which will work with your video recorder will not necessarily be suitable for use with your computer, nor can you use a TV receiver with monitor inputs unless a suitable interface card is present.

When you look at a TV display, you see a picture which is an illusion that is caused by the way that the human eye works. Though animals can sometimes respond to TV pictures, it is almost certain that they do not see the images that we see and are responding only to an impression of movement. All that actually appears on the screen is a dot of light which moves across and down the screen, varying in brightness as it goes. This movement is called *scanning*, and by making the dot return rapidly to its starting side of the screen after each scan across or down, the form of the scan is a single line. By making the horizontal speed of scanning much greater than the vertical speed, the form of the scan is a set of almost-horizontal lines that extend down the screen, Figure 4.1. This is the type of screen display that was envisaged by the true inventor of TV as we know it, Campbell Swinton, whose description is dated 1907.

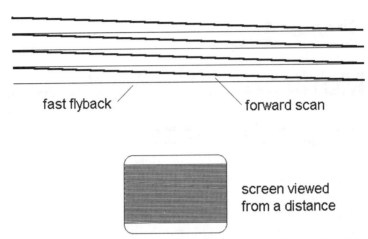

fast flyback   forward scan

screen viewed
from a distance

Figure 4.1 *Scanning consists of moving the electron beam across the tube face at a steady speed, combined with a much slower movement downwards. At the end of a horizontal scan, the beam is returned very quickly. This traces out a pattern of almost-horizontal lines called a raster. The scan angle has been exaggerated here to show the principles.*

The resolution of a TV type of display means the amount of fine detail that can be displayed, and for TV uses resolution is often quoted in terms of the number of vertical lines per screen-width that can be clearly seen. Most domestic TV receivers are rather poor in this respect, even when new and well-adjusted, and a resolution of only 300 lines is not unusual. This restriction is caused by the transmission system rather than by the ability of the cathode-ray tube to display the images, though in earlier days the tube was as much of a limitation. As always, a good monochrome system can usually permit better resolution, but the addition of colour is such a bonus to TV pictures that the loss of resolution is tolerable.

A monitor of professional quality used in a TV studio can produce pictures of quite staggeringly improved resolution, because such monitors can be connected directly to the TV cameras. The monitors that are used with computers are more closely allied to TV studio monitors (with some exceptions as we shall see later), and the resolution is also quoted in terms of dots per screen width, meaning the most closely-spaced dots that can be distinguished on the screen. This is decided by the cathode-ray tube and the electronics circuits that control it, and it requires the dot of light that makes up all of the images to be very small. Some makers of monitors quote the average dot size in millimetres, but this is only useful as a way of comparing monitors of the same size. For example, a dot size of 0.50

mm is undoubtedly good on a 16" monitor, but it would look rather less impressive on a 5" monitor. The number of dots per screen width is a much better guide.

The reason that a mono monitor can always look clearer is that its dot is a single dot, whereas colour monitors need to use three dots, one for each of the primary colours of red, green and blue (Figure 4.2). These are the primary colours of light incidentally – the primary colours of paint are the colours that the paint reflects. No matter how well a colour monitor is constructed it must make use of three dots (in practice, short stripes) to represent each point in an image, making its resolution inevitably lower than a monochrome monitor that is constructed to the same standards. This is why good-quality colour monitors are so expensive in the larger sizes, some £800 or more, in comparison to the £100 or so that will buy a monochrome monitor of excellent performance. A few makers of monitors quote the dot pitch, meaning the distance between the colour dots in a set. This is not particularly useful unless you can translate it into final dot size; for example a 0.31 mm pitch corresponds roughly to a 0.58 mm final dot size. Another complication is that the dot size on a monitor is not constant, it varies with intensity (bright dots are larger than dim ones) and with the position on the screen (the dot size is greater at the edge than in the centre). In general, a dot pitch of more than 0.32 mm on a 14" screen is suited to resolutions of 640 x 480 or less and you need a pitch of 0.28 or less for resolutions up to 1024 x 768. Note that monitors define screen size in the same way as TV receivers, measured along the screen diagonal. A 14" screen of the modern flat section will look noticeably larger than one of the older curved shape.

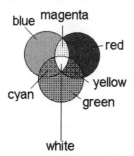

Figure 4.2 *How three dots can be used to produce a range of colours. The dots on the screen do not overlap, and the colours are produced by controlling the brightness of the three close-spaced dots.*

If you aim to keep updating, and particularly if you envisage intensive use of multimedia and Windows 95, you might consider a 17" monitor. These have been prohibitively expensive in the past, around £600, but prices are now dropping, and they do make a very considerable difference to working with multiple window displays.

The CRT

The conventional cathode ray tube (CRT) is cheap, but bulky, heavy, and power-consuming. Readers with an electronics background will be familiar with the principles, but not necessarily in detail if they have specialized in topics other than television. If your previous experience has been in computing rather than in electronics, or if you have little experience of either, the following description may be helpful. If you are familiar with the topics, you can skip this section.

The principles of the mono cathode ray tube (CRT) are shown in Figure 4.3. The screen is coated with a phosphor, meaning a material that glows when it is struck by particles such as electrons or by radiation like ultraviolet. There is no phosphorus involved; both names come from the same (Latin) root meaning glowing in the dark.

At the other end of the tube, the cathode that gives the tube its name, the tiny particles called electrons are torn away from atoms by a high temperature, and accelerated towards the screen by a high voltage, typically 7 – 10 thousand volts (to put this in perspective, the mains supply in the UK is 240V). The stream of electrons is forced into a beam by the metal cylinders in the electron gun, and is focused so that the diameter of the beam is a minimum where it hits the screen, forming the dot of

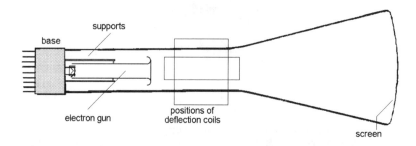

Figure 4.3 *A monochrome cathode ray tube, showing the main features. The deflection coils are external.*

light that appears on the screen. The beam of electrons is moved across and down the screen by a set of coils that fit over the neck of the tube. By passing current through a coil, the coil becomes magnetized and this will shift the position of the electron beam. By continually changing the current through a deflection coil of this type, the beam will be forced to scan across the screen at a steady rate, and by suddenly returning the current to zero, the beam can be made to return rapidly. The use of two coils positioned at 90° allows scanning in both the horizontal and vertical directions.

A colour display needs a much more elaborate type of tube, based on the original shadow-mask type of tube that was first produced by Radio Corporation of America in 1951. The principle of the modern version is that three different phosphors are used and are applied as sets of thin separate stripes down the inner surface of the screen. One phosphor will glow blue, another red, the third green when struck by electrons, and the phosphors are laid in this striped pattern with the three always in the same order and with uniform spacing – there is usually a thin dark band separating the stripes. Instead of one electron gun, the colour tube uses three, and all three beams are focused and moved together. Near the screen, however, the beams have to pass through a set of slits in a stainless-steel plate, called the shadow-mask, and this mask is the point where the beams intersect. When a set of beams has passed through a slit in the shadow-mask (Figure 4.4), the individual beams, which come from slightly different directions, will hit the phosphor stripe of the correct

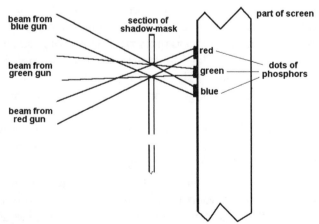

Figure 4.4 *Principles of a colour display. Three electron guns are used, and the shadow-mask, together with the positioning of the phosphor dots, means that electrons from one gun can strike only dots of one colour.*

colour. The shadow-mask prevents the beam from the red gun from hitting the blue or green phosphors, and similarly prevents the beams from the blue and green guns from hitting the wrong phosphor stripes. The size of the shadow-mask holes or stripes defines the pixel size for a colour tube.

The form of the colour tube means that a lot of the electrons on the beams are wasted because they land on the shadow-mask rather than on the phosphor, and to make a colour tube look reasonably bright much higher voltage levels are needed to accelerate the electrons. Voltages of 14,000 to 22,000 volts are used, and at the higher voltages the amount of X-ray generation in the tube starts to become measurable (meaning that it is almost as much as reaches us naturally from the sun and stars). The use of leadglass for the tube reduces the penetration of such X-rays to a negligible amount, less than we receive from natural sources. The colour tubes used for TV receivers use even higher voltage levels, with considerably more X-ray generation.

Dots and signals

The resolution of a monitor in terms of dots per screen-width is just the start of the specification of a monitor for a PC. Nowadays, few users would want to make use of a monitor with fewer than 640 dots per screen-width resolution, but the number of dots per screen-width is not a complete guide to the suitability of a monitor, because there is an additional factor, the way in which these dots are controlled.

The simplest way of controlling the dot brightness on a monitor is simply to turn the dot on or off, so that 'on' means bright and 'off' means dark. This scheme is used for many mono monitors that are intended for mono display boards such as Hercules (see later), and is sometimes described as digital or TTL. The letters TTL refer to a family of silicon chips that make use of two levels of voltage, 0 and 5V, for their input and output signals.

Digital signals are well suited to text, and can be used for a range of graphics shapes as well, so long as you do not aspire to delicately shaded pictures. For colour displays, these TTL signals can be applied to each of the colour primary signals, allowing a limited range of colours to be produced by using single colours or mixtures. Signals that use the three colour primaries in this way can produce red, green and blue using the dots for primary colours and can also show the mixture of red and green, which is yellow, the mixture of red and blue, which is magenta, and the mixture of blue and green, which is called cyan. The mixture of all three colours is white. Using a simple on-off scheme like this can therefore produce eight colours (counting black and white as colours).

The simple on-off method can be improved by using a fourth on-off signal, called brightness or luminance, whose effect when switched on is to make the colours brighter when this signal is switched on, so that you have black and grey, red and bright red, and so on, a range of 16 colours including black. This 16-colour system has been so common in the past that a lot of software still features 16 colours despite using a graphics system that allows a much greater range of colours.

The alternative to digital signals is to use the analogue type of signal that is used for TV monitors, in which each signal for a colour can take any of a range of sizes or amplitudes. This method allows you to create any colour, natural or unnatural, by controlling the brightness of each of the primary colours individually. A monitor like this is much closer to TV monitors in design, and some (but certainly not all) monitors of this type can be used to display TV pictures from sources such as TV, camcorders and video recorders.

There are two varieties of analogue monitors, but the VGA type of monitor uses what is called the RGB set of signals. There is a separate connection for the three colour signals and for the other (synchronizing) signals as well, making the connector require more pins than would be needed for the other type, the composite signal type.

The alternative is a mixed signal which combines the separate RGB components into a single signal, as is done for transmitted TV signals, and is the type of signal that a video recorder will deliver at its video output socket. This means a simpler connection, with only one pin on the plug rather than three or four, but it requires the monitor to separate the signals again, and the resolution will suffer slightly because of this extra processing. On the other hand, such a monitor could double as a monitor for a video recorder or camcorder – but that depends on the scanning system that is used as well. Composite signal monitors are not used for the PC.

This leads to the third point about monitors. All monitors use the line scan system, but for computing purposes this does not have to be identical to the method that is used for TV. Television has always used a form of scanning called interlacing, Figure 4.5, in which the odd-numbered lines (1, 3, 5, 7 etc.) are scanned, leaving a gap between the lines, for one screenful or field, and the even-numbered lines are then

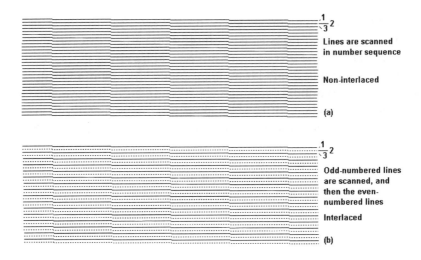

Figure 4.5 *Interlaced and non-interlaced scans. Interlacing is used on TV signals to ease transmission problems, but it is not necessary for computer monitors to follow this system.*

scanned, filling in the gaps, on a second field. Two such fields make up a complete picture frame.

This system was adopted when the TV system as we know it was first being designed in detail around 1935, and the reason was that it made the TV signals easier to transmit with the technology that was available at the time. The original rates of scanning that were chosen for TV allowed for the use of 405 lines at 50 fields per second in the UK, 525 lines at 60 fields per second in the USA. The UK and the rest of Europe then changed over to a scanning system of 625 lines and 50 fields in the mid-sixties, along with the colour TV system (called PAL) which was adopted by most of western Europe, apart (as usual) from France which used its own SECAM colour system and for some time used an 819-line scanning system. The American continent along with Japan stuck with the 525-line standard, now some 54 years old and beginning to look its age. US receivers also use the 1952-vintage NTSC colour system, and these differences in scanning rates and colour systems explain why you can't exchange video-cassettes with friends in other countries.

For computing purposes, there is no need to be tied to TV scan rates, or interlacing, and the different TV colour systems refer only to the way that colour signals are transmitted, not to closed systems using monitors.

In the UK, the 625-line TV picture is achieved by using a horizontal scanning signal that repeats at about 15,000 times per second. This is described as a 15 kHz scan, with k meaning kilo (one thousand when applied to electronics) and Hz meaning Hertz, the unit of repetition rate in terms of the number of repetitions per second. Some computer video systems use the TV rate of 15 kHz, others achieve better resolution in the vertical direction by making use of higher rates such as 22 kHz. The rate of field repetition is usually 60 rather than the European 50, though most monitors allow for adjustment to either rate. The important point is that your monitor must be suited to the computer – the computer decides on the scan rate and the monitor must be capable of using this rate. If your computer uses a 15 kHz scan rate then buying an expensive monitor which can use a 22 kHz or higher scan rate will not improve the resolution – either the monitor will not work, or it will work only at the scan rate of the computer: more correctly, at the scan rate of the graphics board in the computer.

There is also a third requirement. The video signals are used to alter the intensity and/or colour of the dot of light as it moves across the screen, and the dot has to be in the correct place on the screen for each part of the signal. This requires the signals to be synchronized to the scanning, so that the first dot signals for a line arrive just as the scanning of the line has started. If this synchronization is not perfect then the picture will, at best, be mis-shaped, and at worst completely broken up and unrecognizable.

Synchronization is done by another set of signals, the *sync* signals, which are used to force the monitor to start a scan. There are two sync signals, one for each scan, with the line sync signals arriving at the line scan rate (15 kHz, 22 kHz or whatever standard is being used) and the field sync signals arriving at the field rate, usually 60 Hz. These signals can be taken along another two wires in the video cable to two more pins on the plug, or they can be combined with the video signals to make a composite video signal as used in television. Computer monitors for a PC machine work with separate colour and sync signals, but monitors for other computer types use a variety of methods, some using combined colour and sync signal inputs.

The unfortunate point is that computer manufacturers, unlike TV manufacturers, have no agreed standards. Some computers send out positive sync pulses, meaning that the voltage on the pin suddenly rises and very rapidly falls back to zero. Others use negative pulses, with the normally-high voltage suddenly reducing and then returning almost immediately to normal. Some may even use positive pulses for one sync and negative for the other. Unless the monitor can be switched, preferably automatically, to recognize the correct pulse direction (or polarity), synchronization will be impossible.

This lack of standardization has led to the design of monitors which can cope with different scan rates and sync pulse polarities. These monitors are called *multisync*, using a name originally devised by the NEC corporation when they made the first monitor of this type. Modern multisync monitors will adapt, usually automatically, to the scan rate and sync pulse polarity which they detect at the input, and will display a picture from most types of computer signals. This applies mainly to PC machines, however, and some of the mainly-games machines which are still around demand very specialized monitors. Since these machine are unlikely to be used for serious purposes (other than DTP work), we can leave them out of consideration here.

Another point which causes a lot of confusion is the description of monitors as interlaced or non-interlaced. Whether or not a picture uses interlacing is determined by the synchronizing pulses that are sent out from the computer, not by the monitor itself. A monitor that is described as *non-interlaced* means one which will accept the high-resolution non-interlaced signals that are sent from graphics cards that use resolutions greater than 800 x 600. Higher resolution signals in turn demand that the monitor circuits can handle a much larger range of frequencies (the bandwidth), with bandwidths of more than 25 MHz being typical. As a comparison, the bandwidth of a good colour TV receiver is around 5.5 MHz. We should really refer to these monitors as wide-bandwidth rather than non-interlaced.

Video Graphics Cards

When the first IBM PC machine appeared in 1982, its provision for display was a simple black-and-white (monochrome) monitor of 12" diameter. Inside the computer, the numbers that are stored in the memory of the machine were turned into signals suitable for the monitor (video signals) by a separate circuit on a card that was inserted into one of the spare slots of the computer. The design of the machine allowed for adding on all kinds of extra facilities by way of cards which could be plugged into these expansion slots, and since the early machines provided almost nothing in the way of the facilities that we now take for granted (such as connecting printers, floppy disks, extra memory and so on) these slots were a very valuable way of upgrading the machine. Even today, when most PC machines come very fully equipped, a set of four to six vacant slots is still a very valuable part of the specification.

The original type of display was concerned only with text, because the concept of a machine for business use at that time was that only text, along with a limited range of additional symbols, was all that was needed for serious use, as distinct from games. The video card was referred to as the Monochrome Display Adapter, a good summary of its intentions and uses, and usually abbreviated to MDA. It produced an excellent display of text, with each character built up on a 14 x 9 grid of the type illustrated in Figure 4.6. The text was of 80 characters per line and 25 lines per screen, and at a time when many small computers displayed only 40 characters per line using a 9 x 8 grid, this made text on the IBM monitor look notably crisp and clear. Typical monitors used an 18.4 kHz horizontal scan rate and 1000 lines resolution at the centre of the screen.

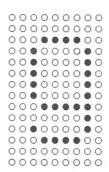

Figure 4.6 *How the shape of a character is displayed on a high-resolution grid or matrix.*

A good monitor with the MDA video card still looks good, and unless you have a special need for graphics and colour displays (which is less usual than you would imagine) then a mono video card and mono display (which can be amber or green rather than white) is considerably easier on the eyes, particularly for text uses like word processing, accounts, spreadsheets and other business applications. The only real fault, as far as text display was concerned, in the original IBM monitor display was that the monitor used a type of cathode ray tube which is classed as high-persistence, meaning that the display faded out rather slowly, so that when one piece of text was replaced by another, the old version would fade slowly away rather than disappearing at once. This, however, makes the display very steady, with no trace of the flicker which many users find very disturbing, and most mono monitors are made with fairly long-persistence screens.

The other fault, however, was that graphics could not be displayed, so that graphs could not be produced from spreadsheets (on screen at least), and applications such as desk-top publishing (DTP) or computer aided design (CAD) were out of the question. This latter point was purely academic at the time, however, because DTP did not exist at that time, and CAD was at an early stage of development on much larger machines. For a modern machine, however, no-one would consider the MDA card or any other card that did not supply full graphics capabilities either in colour or in monochrome.

The problem with the early IBM PC was that IBM did not really know what the main market for the machine would be. At that time, many small computers were aimed at the hobby user, and their applications were mainly games of the zap-the-alien variety, the type of program that I have always thought of as haemorrhoids in space, a pain in the eyes. To understand what the earlier cards did and why they should now be avoided we have to dip into history for a moment, but these cards are still on offer, and might look quite a bargain until you know what their limitations are.

The old graphics cards

The IBM PC was originally conceived as a home computer, but its text video adapter could not possibly produce the graphics that were essential for games, and certainly not in colour. The result was a step that, though it seemed logical at the time, turned out to be retrograde and a source of confusion for years, the Colour Graphics Adapter or CGA card. The CGA card was an attempt to provide some colour graphics capabilities, but it was not well matched to the capabilities of the machine, and certainly did not provide as good a display as most games machines of the time. The card allowed two forms of screen display, a text screen for the display of text characters only, and a set of graphics screens with different resolution capabilities. The board was intended for a maximum resolution of 640 dots across by 200 down the screen, and only when two colours were being used, corresponding to a monochrome display. In its very limited four-colour display (and these four include black and white), the resolution was only 320 dots across by 200 down.

Even when colour was ignored, the CGA card was quite unsuited for either serious or games use, because its text characters were formed from an 8 x 8 set of dots rather than the superb 14 x 9 of MDA. This made the characters look blurred, coarse and poorly-shaped, and quite unworthy of the machine, and even when a CGA card is used in a graphics mode, the resolution is still low. CGA boards are never used on machines intended for serious business uses, and low-cost machines which are aimed at the home user or the smaller business user almost invariably feature monochrome cards. This has not prevented several major manu-facturers from offering machines with CGA cards, and such machines are acceptable only if the CGA cards can be removed and replaced with a better type. Several older portable machines used CGA screens, and these should be avoided, even at bargain prices, because the circuits are built-in rather than being on a replaceable board.

CGA cards require a monitor which uses a 15 kHz horizontal scan rate and digital RGB signals. A few mono monitors can make use of the RGB signals, but there is little advantage in using monochrome because the resolution of the CGA card is so poor. On most computers, a CGA card can be replaced with a more suitable display card, but some old ma-chines, notably the Amstrad PC 1512 and the Olivetti Prodest PC2, made no provision for a separate display card, with CGA built in to the main board.

The CGA card brought one important new principle, the use of 'screens', meaning different resolution modes that could be switched by software. All subsequent cards have featured a text mode and several graphics modes that can use various resolutions and choices of colours. The next step was made, however, by a card that did not offer colour. In 1983, the independent firm of Hercules Inc. designed and marketed a display card which could provide both the original 14 x 9 characters of the original MDA card, and also allow high-resolution graphics display in monochrome on the original type of IBM monitor. This point is important, because the use of the original monitor made the change to a Hercules card one that involved the minimum of cost and effort. The original Hercules video card allows graphics to be drawn with up to 750 dots in each line, and up to 348 lines on screen. This gets the best of both worlds, with excellent displays of both text and graphics, though in monochrome only.

Though there have been several designs of later Hercules cards, the name is still used as a guarantee of excellent monochrome text and graphics display, so that many machines now offered for sale specify either Hercules or Hercules-compatible graphics cards. The use of a Hercules or Hercules-compatible mono card means that you must use a suitable monitor, a mono digital type. A few Hercules-compatible cards allow other display systems to be selected so that other monitors (such as analogue types) can be used.

The Hercules card, excellent as it was, could not have made much impact if the manufacturers of software had not written programs that could make use of Hercules graphics. After all, programs which relied entirely on text could continue to use the older MDA type of card and no changes were needed to run such programs with the Hercules card. The breakthrough for Hercules came when the newly-devised Lotus 1-2-3 spreadsheet was rewritten to make use of the Hercules card for monochrome graphics, along with MDA for text-only and CGA for colour graphics. This allowed 1-2-3 to display business graphics in good-quality monochrome or poor-quality colour, and the Hercules displays were so good that software houses from then on established the Hercules card as their standard for monochrome graphics.

EGA and onwards

The Hercules card was not the only challenge to established IBM standards, because several manufacturers (notably Compaq, Olivetti and AT&T) were making use of improved CGA cards which allowed 640 x 400 resolution by making the monitor work at double its normal field scan rate. This latter development requires specialized software support as well as specialized monitors and has now died out – beware of such cards offered as a bargain. Eventually the pressure for better display cards forced IBM to offer an alternative to MDA and CGA. Their problem was to maintain compatibility with their other boards, so that software written for MDA or CGA could still be used with the later cards, and this concern with compatibility led to the need for compromises.

The new card was called EGA, enhanced graphics adapter, and it relied on increasing the horizontal scanning rate from the 15 kHz of CGA to around 22 kHz so as to permit a resolution of 640 x 350 dots. In addition, an analogue signal is used, so that the choice of colours is greater. The normal range of colours is 16, but the 16 colours that can be displayed can be selected from a set of 64, so that 16 is the practical limit to the number of colours that can be displayed on one screen, rather than the absolute limit to the number of colours. In addition, the EGA board could display monochrome text on a 14 x 9 matrix, monochrome graphics of 720 x 350 resolution, and also CGA-type colour displays of 320 x 200 resolution. This allowed the EGA board to be used along with software which had been written for the earlier type of graphics displays, though it made no provision for software that catered for the Hercules card. The feeling seems to have been that a 'third-party' product should not be encouraged in this way, even though the Hercules card had been instrumental in expanding the uses of the machine.

Very soon after the EGA board appeared, software suppliers revised programs so that full advantage could be taken of the EGA standard. This started a race between hardware suppliers and software suppliers which has caused some of the enormous confusion that this book is intended to clear. Hardware suppliers have introduced new video graphics boards which, while allowing the use of Hercules mono graphics and all of the IBM standards, have also permitted the use of much higher resolution colour graphics. Software suppliers have tried to adapt their programs so as to cater for all the possible graphics boards. In addition, some suppliers of boards also supply software in the form of drivers which allow their boards to be used to better advantage. This race has lead to the situation that we have today in which a main source of video display problems is incompatibility between the video card and the software.

When IBM introduced their PS/2 machines in 1987, they broke to a considerable extent with the compatibility that had persisted from the early days of the PC machines and extended to the AT machines. The PS/2 machines no longer use a separate video card, a point of construction that was already being used by other manufacturers such as Amstrad. In addition, the PS/2 machines featured two new video display systems, MCGA (Multi-colour Graphics Array) and VGA (Video Graphics Array), and a new range of monitors which were analogue only. This meant that users of monochrome machines had to invest in new monitors, because up until that point monochrome machines had used Hercules-type cards with digital signals into the monitors.

The MCGA type of card is not likely to be widely emulated, and is not compatible with MDA or EGA. The VGA card, however, has set the standard that business software has followed and is being widely emulated by other cards. VGA permits full compatibility with MDA, CGA,

EGA and even MCGA. In addition it adds displays of 640 x 480 16-colour graphics and 720 x 400 colour graphics, using a 9 x 16 grid for characters with colour. This card requires a high standard of monitor if colour is required, and the use of a mono monitor can be restricted to a few modes unless an analogue monitor is used. Analogue mono monitors, often referred to as VGA mono monitors, are less common now than they were a few years ago, and it is important to check if you want to use a mono monitor with a VGA card. If an analogue colour monitor is to be used, it should have a resolution that matches that of the graphics card. Because modern VGA cards allow earlier standards to be used if required, it is a false economy to fit anything other than a VGA type of card to a modern PC machine.

You should not consider using a CGA or EGA card in any machine that you build, no matter how attractive the price. If you are really strapped for cash, consider using a Hercules mono graphics card along with a TTL mono monitor, which will produce crisp and clear mono-chrome pictures using 720 x 348 resolution. Unless lowest possible cost is essential, use a VGA card and a suitable matching monitor. If you buy a mono monitor that allows the use of SVGA (allowing 800 x 600 graphics displays) you will eventually be glad of your foresight.

Enhancements and confusions

The VGA standard is so good that it has become a standard fitting, and no other video card is likely to be found on a modern machine. As a result dozens of suppliers have vied with each other to design and manufacture VGA cards. Some of the respected names in the video card business include ATI, Genoa, Orchid, Paradise and Video Seven, and many well-known brands of computers will be found to include video cards by one of these suppliers. In addition, there are many less well-known suppliers who use these cards or cards by equally less well-known manufacturers. As long as these cards are genuinely compatible with the IBM standard then there should be no problems with software.

The trouble comes mainly if you want to use enhanced features. You can find 640 x 480 resolution in 256 colours or even in 16 million colours, 800 x 600 in 16 colours and even 1048 x 768 or more in four colours or sixteen colours. There is a trade-off between resolution and number of colours, because increasing either requires more memory for video, and the amount built into the video card is limited, often to 256 Kbyte. Unless the card contains additional memory sockets that allow you to upgrade its memory, the higher resolutions must inevitably be restricted to smaller numbers of colours at any given time (though the colours used on a screen can be picked from a large range). The number of colours corresponds to the number of digital bits used to code colour, and it does not imply that your monitor can display such a range of colour or your eye detect any differences between them. A memory of 512 Kbyte is adequate for most purposes unless you want to use more than 256 colours.

Breaking away from the dominance of IBM in setting standards, a number of manufacturers of graphics cards have co-operated in setting a common standard called VESA (Video Extended Standards Association) so that producers of software can write drivers for the standard VESA resolutions. VESA supports the 800 x 600 16-colour mode, and this is the most important of the enhancements to VGA. Its resolution implies the use of square dots on the screen and is attainable by the better multisync monitors. The aim is that software producers can use a software driver which will suit any board produced by a VESA member to allow the use of all the VGA modes and the super-VGA mode of 800 x 600. The 800 x 600 mode is also found in most of the VGA-type cards which are built into new computers as well as in add-on cards for older machines.

If you are tempted to use higher resolution graphics it is important to remember that you will need a monitor that is capable of displaying the higher resolution. This will usually be a non-interlaced type of monitor. You will also need a graphics card with at least 512 Kbyte of memory, possible up to 1 Mbyte. Having done this, you may find the displays of words and icons so small that you need to work very close to the screen. This is undesirable, so that high resolution should be used only along with larger screens, and for purposes like image editing and DTP work for which it is desirable. This in turn makes it very expensive.

Choosing a graphics board

The PC is unusable unless a video graphics board of some sort is fitted, and following the advice noted earlier you are likely to want to use a VGA card of one variety or another. A quick flip through the computing magazines reveals that this is not a simple choice, because most suppliers can provide a considerable number of video cards at a huge range of prices, and monitors that follow similar patterns.

Here, as always, you need to think of what you need to use the computer for, and the trouble is that unless you have had some experience with a PC machine it is not easy to make an informed choice at this point. One item that can decide you is the general type of programs that you will use. These are grouped as DOS or Windows programs, and the important difference is that Windows programs make intensive use of the high-resolution graphics ability of the machine. If all or most of your programs make use of the Windows system, then you need a graphics card that is fast-working, and if you intend to use resolutions higher than 640 x 480 you will need a video graphics card with more than the usual 256 Kbyte of memory.

On the other hand, if your programs make use only of DOS, and are primarily text-based (such as word processors), you do not need a fast graphics card with a lot of memory, and you could quite happily use the lowest-priced VGA card along with a mono monitor from Philips or Samsung. Such a combination will also cope with Windows, and this text is currently being typed on a machine using a low-cost VGA card, Philips mono monitor, and running the word processor Word for Windows. You cannot, however, run the 800 x 600 video displays on this set-up, and for the purposes for which it is likely to be used this is no loss. If you want to use 800 x 600, it is easy enough to find a low-cost card, but a low-cost monitor is not so easy to find, though several firms are currently advertising SVGA mono monitors for under £100. You can expect to pay more than twice this amount for colour.

Straightforward simple VGA cards can be found that are described as 8-bit. Unless your needs are very modest, running mainly text-based DOS programs, it is better to buy the type that are described as 16-bit. The minimum video card memory is 256 Kbyte, and if you will want to use 800 x 600 resolution you should specify 512 Kbyte. At the time of writing the price difference between the extremes of 8-bit 256 Kbyte and 16-bit 512 Kbyte was only around £10, making it pointless to go for the lowest-cost option.

If you do not intend to make use of high-resolution displays at present, it can still be an option for the future if the video card allows more memory to be plugged in. Video cards are likely to use separate memory chips rather than SIMMs, and you need to know which chips have to be used. If the video card has no spare memory sockets, no further memory can be added and if you subsequently want to use high resolution displays you will need to change the memory card for a new one.

Another video card option is speed. The time needed to put a display on the screen can limit the speed of the PC machine when you are using Windows (see Chapter 9), so that several types of video cards have been developed which work faster at this task than others. Like Hi-fi, you can pay a lot of money for fairly modest increases in speed, and for many purposes, the cards which use the Tseng Laboratories set of chips are gratifyingly fast without causing too much disruption to your bank account. A card of this type should cost less than £100, but for a little more than £100 you can find cards from Orchid that are noticeably faster. For very high speeds, cards that use a local bus and cost around £350 offer the fastest performance obtainable. These very fast cards also use expensive memory chips, and they are justified only if you need the utmost in fast performance. They must be fitted to a local bus, and since you would use such performance to match the speed of a Pentium, the PCI bus is the preferred option. In addition, the faster video cards will use jumpers to select various combinations of characteristics; a typical set would be vertical frequency figures of 60, 70, 72 and 87 frames per second for 1024 x 768 and 256 colours, or rates of 56, 60 or 72 frames per second for 800 x 600 and 256 colours.

Connectors

The standard connector for VGA is a 15-pin D type, whose shape is shown in Figure 4.7.

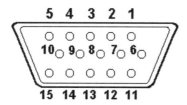

Figure 4.7 *The standard 15-pin connector for a VGA monitor.*

The pin use is given overleaf.

When the monitor is a mono one, only the green video signal is used. This needs to be remembered if you are working with colours as shades of grey, because it means that pure blue and pure red will display as black, and only colours with a green content will display. If you are using

15-pin connector	
Pin	**Use**
1	Red out
2	Green out
3	Blue out
4	NC
5	Earth
6	Red earth
7	Green earth
8	Blue earth
9	No pin
10	Sync earth
11	NC
12	NC
13	Horizontal sync
14	Vertical sync
15	NC
Notes:	NC = No connection, pin not used. Pin 9 is removed to act as a key.

software that permits you to construct a 'palette', a range of colour signals, make sure that each colour you specify in terms of green, red and blue has a sufficient green content to be visible on a mono monitor.

Some monitors that you may be offered have a 9-pin connector. These monitors may be used if they offer the correct signal connections, but you

9-pin connector	
Pin	**Use**
1	Red out
2	Green out
3	Blue out
4	Horizontal sync
5	Vertical sync
6	Red earth
7	Green earth
8	Blue earth
9	Sync earth

need to be absolutely certain that they will take the correct polarity and amplitude of signals. Monitors that are suitable for the PC and which use a 9-pin connector are likely to use the following pin-out, and you can either connect the cable to a 15-pin plug or make up a 9-to-15 adapter.

Remember that a monitor with a 9-pin connector may be a TTL mono type, and intended to be driven by a Hercules card. If you have some circuit experience with monitors, you might be able to adapt a TTL monitor for analogue use by altering the bias on the video amplifier stage(s). The standard video input amplitude for an analogue signal is 0.7V peak-to-peak, so that you might need to add another video stage. On the whole, such a conversion is undesirable unless you are knowledgeable, experienced and keen to experiment.

Drivers

For all normal purposes, your software will cope with a VGA card and its matching monitor without any intervention from you, providing that you are using standard text mode or 640 x 480 resolution graphics. When software is installed, you will be able to specify what video card and monitor (colour or mono) you are using, and by selecting VGA as the video card type, the system will automatically be configured to send out the correct signals. If you have a mono monitor, it is better to select the mono option, though you can use the colour option if you want. The difference is that selecting the mono option avoids the use of pure blue or green signals, since the monitor uses only the green signal. For some software, notably Windows (see Chapter 9), it can be an advantage to select Colour even if you are using a mono monitor, because this option allows you more control over the screen appearance than the mono option, assuming that your monitor is a VGA type. If you are using mono Hercules, always take the mono option as your monitor type.

Complications arise if you want to use SVGA resolutions, typically 800 x 600. Contrary to what is sometimes suggested, you can run 800 x 600 even on a card with only 256 Kbyte of memory if you keep to a small range of colours, and the main limiting factor is whether your monitor will permit the use of the different sync rates, typically 35.5 kHz and 57 Hz for horizontal and vertical respectively. Most mono monitors use a fixed horizontal frequency of 31.467 kHz and cannot be used with 800 x 600 displays. Modern colour monitors (and many mono types) are likely to be of the multisync type, which will permit the change of frequency. The amount of memory in the SVGA board determines how many colours can be used in a picture.

What makes it rather less simple is the provision of drivers, meaning the software packages that adapt your programs to the resolution that the video card is using. Video cards generally come packaged with a disk that contains a set of these drivers, and some are of abysmal quality as well as being difficult to install. If you want to experiment with the higher resolution modes, and have suitable hardware, then the system outline here is a useful guide.

First of all, for the programs you intend to use, check to see if the installation or set-up program allows you to specify the use of 800 x 600 or whatever resolution you want to use. If this option appears, then the software contains its own driver, and this is likely to be better suited. Programs such as graphics packages often offer a large range of drivers and when you run programs using DOS you can select to use a different driver for each program. Users of Windows will find a large selection of drivers for VGA and SVGA modes.

If your programs offer no high-resolution options you will need to make use of the drivers that have been supplied along with the video card. This usually involves running a small program prior to the main one, and the process can be automated by using a batch file (see Chapter 9). The manual that comes with the video card will list the programs for which drivers are available, and it is at this stage that you are likely to find problems, either that the programs for which drivers are available are not the programs you want to use, or that the drivers are written for very old versions of these programs. Unfortunately, a lot of video cards on special offer are of ancient design and include ancient drivers.

If you intend to use Windows for all your programs, the task is much easier because you need install only one high-resolution driver for Windows itself. You need to check that the Windows driver of your video card is suitable for the up-to-date Windows version. A few video cards still refer to Windows Version 2 which is completely obsolete now, and only drivers for Windows 3, preferably Version 3.1, can be used. Before you install such a driver from the disk that comes with the video card, try some of the built-in Windows high-resolution drivers, as they are likely to be better. Use the SETUP program of Windows to install the driver, which usually involves inserting three of the Windows distribution disks (numbers 1, 2 and 5) to read files from these disks.

Chapter 5

Ports

Port cards are intended to carry out the interfacing actions that are needed in order to connect a PC to other devices. Parallel port cards are used to connect the computer to printers, and can also be used for other devices which need fast transfer of data, typically additional disk drives and tape streamers. A parallel port deals with a byte of data (8 bits) in the transfer, using eight data lines. Serial ports are used for modems and for linking PC computers together into simple networks. These work one bit at a time, and at their simplest need only a single connection (and earth return) between the devices that are connected. Serial ports are seldom quite so simple, however.

Mouse ports are more specialized, used only for the mouse or its equivalent trackerball or graphics digitizer, and when you construct your own machine it is most unlikely that you will be able to lay your hands on a mouse port card unless you buy a mouse that has the card packaged with it. Since most computers have at least one serial port, it's usually cheaper to buy a serial mouse that uses the serial port. The keyboard port is permanently fitted to the motherboard as distinct from being added in the form of a card .

For anyone constructing a machine from scratch and using the standard form of IDE card for the disk drives, the provision of port cards is not a priority, because most IDE cards come with ports attached, usually one parallel and one or two serial. If, however, you have one of the motherboards that includes the IDE interface on the motherboard itself, you will (usually) need to install a separate floppy drive card, and unless this card includes port actions you will also need to put in parallel and serial port cards. The most useful option is a multi-function card that offers a parallel port and two serial ports, because if you opt for just one serial port it is likely that you will want to use another one sooner or later. Having only one parallel port is much less of a restriction.

The provision of one parallel and two serial ports, particularly if the IDE board includes one parallel and one serial port, is simple and unlikely to cause problems. If, however, you want to use two parallel ports or three serial ports, or if there are ports provided on more than one card, problems can arise because of conflicts of addresses and interrupts.

The address for a port card means the code reference number that is used to locate the circuits on the card. By using this method, data can be sent to a port using much the same methods as when it is written to the memory by using a memory address number. A port needs to be able to use a unique address number because if two ports shared the same address there would have to be some additional method of determining which one is to be used.

The interrupt is a signal that the computer uses as a signal to divert its attention. If your computer is busy attending to the keys that you are pressing, and a port needs to input data, an interrupt signal from the port will cause the computer to suspend servicing the keyboard and turn its attention (by running a different routine) to servicing the port, reading in or writing out the data. When this has been done, the earlier routine is resumed where it left off, and the break is often quite undetected. An interrupt is the way in which a port can make use of the computer without the need to shut down programs. Once again, conflicts of interrupts can cause port problems that can be difficult to resolve.

The signals that are used to synchronize the computer to the needs of whatever is connected to the port are called 'handshaking' signals. A typical handshaking sequence might consist of a computer sending out the 'Are you ready?' signal through the port to the printer or other device. This will continue until the other device sends back a 'Yes, ready' signal. Data will then be sent until the remote device sends a 'Hang on' signal, and the flow of data will be halted until this signal is removed. The word 'handshaking' emphasizes the interaction that is needed, requiring more than one transmission of control signals between computer and remote device. Another term for the same sort of thing is 'flow control' and you are likely to meet with both of these phrases when you deal with communications hardware and software.

The parallel port – principles

The parallel port is often termed a Centronics port because its standardized format is due to the printer manufacturers Centronics, who devised this form of port in the 1970s. Centronics port sockets on most small computers use a 36-pin Amphenol-type of connector, but the PC

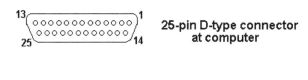

25-pin D-type connector
at computer

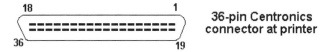

36-pin Centronics
connector at printer

connectors viewed into pins

Figure 5.1 *Printer connectors - the connector at the PC is a 25-pin D-type, but a standard Centronics type is used at the printer. Some printers allow for serial connections and use a 25-pin D socket for this.*

uses a 25-pin D-type female connector at the PC end of the cable, and the 36-pin Amphenol-type at the printer end.

These connectors are illustrated in Figure 5.1, and the corresponding pins are:

Signal	DB-25	36-pin
Strobe	1	1
D0	2	2
D1	3	3
D2	4	4
D3	5	5
D4	6	6
D5	7	7
D6	8	8
D7	9	9
Ackn	10	10
Busy	11	11
PE	12	12
Slct	13	13
Auto	14	14
Error	15	32
Reset/Init	16	31
Slct In	17	36
Earth	18-25	16,19-30,33
NC	-----	15,18,34

The D0 to D7 signals are the eight data lines, and the direction is always out *from* the computer. When a parallel connector is used for two-way data communication, four of the data lines are used for the outward signals and four of the control pins are used for the input signals. This makes the use of a parallel port for two-way communication rather slower than the use of the signals on the slots, so you should use a parallel-port disk drive, for example, only if the installation of an internal drive is impossible.

In normal use, each data pin has its own earth pin, and the connecting wires are a twisted pair for each, so minimizing interference between lines. The control signals are as follows, taking an input as meaning a signal into the computer from the printer, and an output as a signal from the computer to the printer. Note that printer manuals use the opposite sense for input and output.

- *Strobe* is an active-low pulse output to the printer to gate the flow of a set of data signals on the D0 – D7 lines.

- *Auto*, if used, is an output to the printer that, when held low, ensures that the paper is fed on after printing a line.

- The *Reset/Init* signal output, active low, will cause the printer to be reset, clearing all data that might have been in the buffer memory of the printer.

- The *Slct In* output, also active low, inhibits all output to the printer unless this line is held low. Problems with 'dead' printers are often traced to this latter line being disconnected.

The remaining control signals are all inputs to the computer from the printer.

- The *Ackn* signal, active low, is a short pulse that acknowledges that data has been received at the printer and that the printer is ready for another byte.

- The *Busy* line is active high to indicate that data cannot be sent because the printer cannot accept it.

- The *PE* input indicates that the printer is out of paper, and *Slct* is active high to indicate that the printer is selected for use (the printer is on-line).

- The *Error* input, active low, signals that the printer is off-line, that the paper-end detector has operated, or that the printer is jammed (paper or ribbon, for example) and cannot operate.

Fitting a parallel port card

The parallel port is normally part of the IDE card, but irrespective of whether it is fitted on the IDE card or on a separate card, jumpers will need to be set for the correct address and interrupt. Address numbers are given in hexadecimal scale (see Appendix C), but you are seldom involved in this because the parallel port will normally be set up ready for use in a machine with one parallel port only. Problems arise only when two parallel ports exist. This can happen unintentionally because if the IDE card includes a parallel port and the graphics card also includes a parallel port, both ports will initially be set for the same address and interrupt number. Another possibility is that you have one of each port on the IDE card, and you have added a multi-port card with the intention of using a second serial port, but this card includes a parallel port as well.

The computer software refers to the parallel ports as LPT1, LPT2 and so on, and also to PRN, which normally means LPT1. When only one printer is fitted, it is normal to set the jumpers on the parallel port to the LPT1 settings, which are:

    Address.......0378   Interrupt.......IRQ7

When two printer ports are in use, these are referred to as LPT1 and LPT2, and the LPT2 port uses the settings of:

    Address.......0278   Interrupt.......IRQ5

In the unlikely event of using three printer ports, the third port, LPT3, will use the address of 03BC and interrupt IRQ5, with the LPT1 and LPT2 ports sharing IRQ7.

If you use a set of cards that provides you with an unwanted printer port, set the jumpers on this port so that it is disabled. This avoids any problems of incorrect port action. Symptoms of conflicts are when a printer that is connected to LPT1 suddenly responds only to LPT2 when a second port is added, even if the ports have been set to the correct addresses. These conflicts can arise even with only two printer ports, and if you want to run two printers from one PC it is often easier to use a printer-switch (see the Maplin catalogue) rather than installing a second port and setting it up. Note that diagnostic programs (see Chapter 8) can be very helpful in pin-pointing problems, and you will also get a screen report on the port addresses and interrupts when you boot up the computer.

Add-on units that make use of a parallel port usually are of the 'feed-through' variety, which means that the plug that engages with the printer socket carries a socket on its back. The printer can be plugged into this socket and used normally because the operating system can control the port so as to distinguish its dual uses. You should not attempt to stack several such add-ons, however. Once again, if you are using a 486 or higher motherboard or if you intend to upgrade, you should consider using a bi-directional (two-way) parallel port. The ordinary parallel port can be used in both directions, but only for 4 bits at a time, and with the connections used in a non-standard way. The bi-directional type is more useful if you are likely to use any of the new fast devices (such as external hard drives or scanners) that plug into a parallel port.

Serial ports – principles

The serial transfer of data makes use of only one line (plus a ground return) for data, with the data being transmitted one bit at a time at a strictly controlled rate. The standard system is known as RS-232, and has been in use for a considerable time with machines such as teleprinters, so that a lot of features of RS-232 seem pointless as applied to modern equipment. When RS-232 was originally specified, two types of device were used and were classed as Data Terminal Equipment (DTE) and as Data Communications Equipment (DCE). A DTE device can send out or receive serial signals, and is a terminal in the sense that the signals are not routed elsewhere. A DCE device is a half-way house for signals, like a modem which converts serial data signals into tones for communication over telephone lines or converts received tones into digital signals. Because the serial port is so closely associated with the use of a serial mouse the information is also covered here. The use of a modem is covered in Chapter 7.

The original concept of RS-232 was that a DTE device would always be connected to a DCE device, but with the development of microcomputers and their associated printers it is now just as common to require to connect two DTE devices to each other, such as one computer to another computer. This means that the connections in the cable must be changed, as we shall see. The original specification also stipulated that

DTE equipment would use a male connector (plug) and the DCE equipment would use a female (socket), but you are likely to find either gender of connector on either type of device nowadays. The problem of how connectors are wired up is one that we'll come back to several times in this chapter. What started as a simple and standardized system has now grown into total confusion, and this sort of problem occurs all the way along the communications trail, not least in the use of words.

The original cable specification of RS-232 was for a connecting cable of 25 leads, as shown in Figure 5.2. Many of these connections reflect the use of old-fashioned telephone equipment and teleprinters, and very few applications of RS-232 now make use of more than eight lines. The standard connector for PC machines is now the D-type 9-pin (Figure 5.3) but even in this respect standards are widely ignored and some manufacturers use quite different connectors. Worse still, some equipment makes use of the full 25-pin system, but uses the 'spare' pins to carry other signals or even DC supply lines.

If any cables that you need along with serial equipment are supplied along with the equipment you have a better chance of getting things

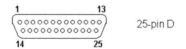

25-pin D

socket as seen when looking
towards the back of the computer.

1*	chassis earth	13	secondary CTS
2*	transmit data	14	secondary TD
3*	receive data	15	transmit clock
4*	request to send	16	secondary RD
5*	clear to send	17	receive clock
6*	data set ready	18	divided clock
7*	signal earth	19	secondary RTS
8	data carrier detect	20*	data terminal ready
9	NC	21	signal quality
10	NC	22	ring indicator
11	NC	23	data rate selector
12	secondary DCD	24	transmit clock extreme
		25	NC

* pins used for serial communications in modern equipment

Figure 5.2 *The old RS-232 form of connector and pin use. The asterisked pin numbers are used nowadays, others are not.*

9-pin D

socket as seen when looking
towards the back of the computer.

1   data carrier detect (DCD)
2   receive data (RX)
3   transmit data (TX)
4   data terminal ready (DTR)
5   signal earth
6   data set ready (DSR)
7   ready to send (RTS)
8   clear to send (CTS)
9   ring indicator (RI)

Figure 5.3 *The modern 9-pin D connector as used on PC machines. Even the pins in this reduced set are not necessarily all used.*

working than if you try to marry up a new piece of equipment with a cable that has been taken from something else. The important point is that you cannot go into a shop and ask for an RS-232 cable, because like the canned foods, RS-232 cables exist in 57 varieties. The advantages of using serial connections, however, outweigh the problems, because when a modem is connected by the RS-232 cable to your computer you can use a simple single line (telephone or radio link), and distance is no problem – wherever you can telephone or send radio messages you can transfer computer data providing that both transmitter and receiver operate to the same standards. A huge variety of adapters (such as gender-changers) can be bought to ensure that your cable can be fitted to a socket that may not be of the correct variety, but this does not guarantee that the connections will be right.

All of this information may look academic, but the conclusion is practical enough. If you are going to join two computers that are in the same room or the same building and pass data between them, you need a serial cable which is described as non-modem, or DTE-to-DTE. If you are going to transfer data over telephone lines or radio links then you need a modem and a modem cable, or DTE-to-DCE cable. The good news is that if you use for your telephone line transfers the type of device that is referred to as an 'internal modem' you don't have to worry at all about this problem of using the correct cable. The differences are that the

Modem connection

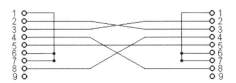

Null-modem or non-modem connection

Figure 5.4 *Connections for normal (modem) and reversed (null-modem or non-modem) serial cables. The 9-pin set is shown, the 25-pin connector has the corresponding pin names connected in the same way. Some applications require minor differences in the null-modem cable.*

modem or DTE-to-DCE cable has corresponding pins connected, and the non-modem or null-modem type has several reversed connections, Figure 5.4.

Setting up the serial port

Whatever you intend to connect to the serial port, you need to have at least one port, and once again, if you are using an IDE card there is likely to be at least one serial port built into this card, with either a 9-pin or 25-pin connector. Note that a 25-pin serial connector is a male type to distinguish it from the 25-pin female connector used for the parallel port of the PC. Many IDE cards provide two serial ports (one 9-pin and one 25-pin), though some are supplied with one working port and the other needing the chips inserted. If you can, specify an IDE card with both serial ports ready for use, because this makes life much easier when (not if) you need the second port.

The serial ports are referred to as COM1, COM2 and so on, but the use of more than two serial ports is fraught with problems because a COM3 port, for example, has to share the IRQ4 interrupt with the COM1 port. This need not be a problem if only one port is used at a time, but if you use the COM1 port for a mouse and the COM3 port for connection to another computer problems will certainly arise, unless you can run software that allows the ports to share the interrupt correctly. This problem arises because the standard PC machine is not well provided with interrupt lines. For most users, however, two serial ports are quite sufficient, and these can be set as COM1 and COM2.

Port number	Address	Interrupt
COM1	03F8	IRQ4*
COM2	02F8	IRQ3*
COM3	03E8	IRQ2
COM4	02E8	IRQ5
		* Standard setting

The use of IRQ2 is not usually possible, and IRQ5 may not be available, depending on the set-up of the machine.

The jumpers on an IDE board that has two serial ports will normally be set to allow the use of COM1 and COM2 and no alterations will be needed. If you add another serial port card, you can set for the address of COM3, but whether or not an interrupt will be available depends on your machine. For example, if you have a second parallel port fitted, IRQ5 will not be available, and IRQ2 is not available if you have a VGA (or EGA) graphics card, which for most of us means never. In addition, if you have added a device such as a scanner which requires the use of an interrupt this will take out any spare interrupt line you had hoped to use.

Once again, it is most unusual to need three serial ports, and software (such as some low-cost networking software) that requires the use of three serial ports is normally supplied with utilities to allow the sharing of an interrupt between two serial ports. The most common use for a serial port nowadays is to make use of a serial mouse, because when a modem (see Chapter 7) is fitted it normally contains its own serial port chip and does not require a cable attachment to a connector on the PC.

The mouse

The mouse, Figure 5.5, is another way in which information is fed into the computer, and for graphics and drawing (CAD) programs, along with many DTP programs, is the only really serious way of working with such programs. A mouse is essential for using Windows. The mouse consists of a heavy metal ball, coated with a synthetic rubber skin, which can be rotated when the mouse is moved. The movement of the ball is transmitted to small rollers, Figure 5.6, and then sensed in various ways,

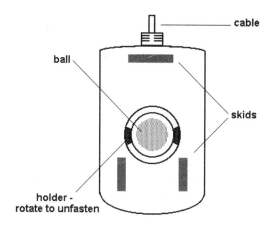

Figure 5.5 *The mouse as seen from its underside. The retainer ring is rotated to release the ball for cleaning.*

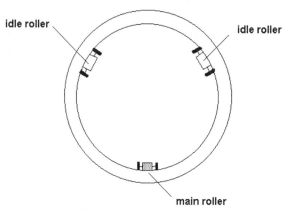

Figure 5.6 *The mouse mechanism as revealed when the ball is removed. The rollers must be kept clean - most mouse problems arise from dirt on the rollers.*

depending on the make of the mouse, such as magnetic changes, or by light reflection, and the signals that indicate the movement are returned to the computer to be used in moving the cursor.

The mouse action depends on software being present, and this can cause a considerable amount of confusion. Typically, the program MOUSE.COM has to be run in order to make the mouse usable with programs that have been designed for mouse action. This does not, however, guarantee that the mouse will work as a cursor mover when you are using your favourite word-processor, spreadsheet or database program. For some machines, like the older Amstrads, where the mouse comes as part of the machine, running MOUSE.COM in the AUTOEXEC.BAT file will allow the mouse to be used as a cursor mover in any program. For other machines, an additional program such as DEFAULT.COM must be run in order to allow this extended mouse action. If you have bought and fitted a mouse independently of the manufacturer of the computer, follow closely the instructions that come with the mouse, and check that the mouse is the correct type for your computer. If you use the Windows system, the mouse action is catered for automatically.

Very few computers are provided with a mouse port on the mother-board, and there are two ways of adding a mouse to such machines. One is to buy a bus mouse, which should be packaged along with an interface card that allows the mouse to be connected to the port on the card. This is, by now, fairly unusual, and it is more common to find that a mouse will be supplied either for serial port connection or for connection to the mouse port, as on the IBM PS/2 machines. The bus-mouse connector for two different computers may look identical, but is not necessarily con-nected up the same way. This may be comparatively harmless if, for example, the left and right movements are reversed (this can be remedied by re-connecting the cable, or making an adapter in which the left and right cables are reversed). On some machines, however, connection of the mouse from another machine could possibly cause damage (though this is unlikely) and should be avoided. Connection of a standard mouse, such as the Microsoft or Logitech mouse, to a PC XT or AT clone with a mouse port or serial port should cause no problems if instructions are followed – it's the mice that are produced for specific models such as Amstrad or Schneider that need to be treated with caution.

The alternative, now considerably more common, to using a bus mouse is to connect the mouse in the way that a modem or another computer might be connected, through the serial port, and a mouse connected in this way is described as a serial mouse. The advantage of using a serial mouse rather than a bus mouse is that it is more likely to be possible to connect a serial mouse to any variety of PC machine, because virtually any PC will have a serial port. The snag is that you may very well need the serial port for other actions, such as the use of a modem or for connections between computers, so that it is an advantage to opt for more than one serial port when you are considering port provisions.

Note that the two mouse types are not interchangeable – the signals that pass between mouse and computer are quite different, and the connectors can be different as well. The bus mouse often uses a 9-pin D connector, and the serial mouse can use either a 9-pin D or a 25-pin D connector. The connections to the 9-pin D-plugs for a serial mouse are not the same as those for a bus mouse, and when a serial mouse is fitted with a 9-pin D connector it is usually packaged with a 9-to-25 pin adapter as well so that it can be used on a serial port which terminates in a 25-pin socket.

The actual physical connection of the mouse to a port does not in itself make the mouse usable, and software needs to be run to activate the mouse. This nowadays is usually a program called MOUSE.COM which is run in the AUTOEXEC.BAT file. This program remains in the memory of the machine until you switch off, so that you do not need to re-activate the mouse each time you change to using another program.

Mouse problems and solutions

Mouse has no effect on cursor
You may be using the wrong type of mouse (serial instead of bus or vice-versa), or the connections might be incorrect. The more usual reason is that the software is not installed. MOUSE.COM must have been run to allow use of mouse in CAD or DTP programs and you may need to run another program with a name such as DEFAULT.COM to make use of the mouse in other, particularly older, programs.

Cursor moves in the wrong direction
The mouse is not intended for the model of computer. Either change the mouse or change over the connections at the mouse plug or make an adapter which has crossed-over connections for the mouse signals. If the reversal is horizontal only, cross over the XA and XB connections. If the reversal is in the vertical direction only cross over the YA and YB connections. Cross over both if both directions are reversed.

Mouse stops working on ordinary programs
after running a program that requires mouse use

This is due to a badly-written section of program which has not restored mouse action correctly after using the mouse, and it particularly applies to old Amstrad machines in which the mouse action is extended to other programs without the need to run an extra driver (other than MOUSE.COM, that is). One well-known desktop publishing program did this, and there may be others. If this happens on a PC which needs to use DEFAULT.COM (or its equivalent) in addition to MOUSE.COM, then the remedy is to run DEFAULT.COM again. A more drastic remedy is to restart the computer (Ctrl-Alt-Del) after running the offending program, but this is the only simple method for the Amstrad machines.

The alternative for Amstrads and any other machines in which MOUSE.COM provides mouse action for ordinary programs is to place the program MARK before MOUSE.COM in the AUTOEXEC.BAT file and put RELEASE at the end of a batch file that runs the offending program, followed by MOUSE. The effect of this is the mark the position of the MOUSE.COM program in the memory and remove it completely after the end of the troublesome program, then re-install MOUSE.COM. This will re-make the connections that have been broken. Simply running MOUSE.COM again is not enough, because MOUSE.COM cannot be run if there is a version, even a damaged version, already in the memory.

Mouse jammed

Clean the ball, avoiding solvents – spectacle lens cleaners and a soft cloth are preferred. Clean also the rollers that you can see inside the mouse casing when the ball is removed. There are often three of these, two of which serve to locate the ball, with a third, spring-loaded, transmitting the movement to a detector. These rollers are quite difficult to clean, and the best way is to wrap some cloth (a handkerchief will do) around the points of tweezers, moisten this, and rub it across each roller, moving it sideways a few times, then pulling or pushing to rotate the roller slightly and then rubbing sideways again. Dirty rollers are a much more common source of trouble than a dirty ball.

Erratic movement

When it becomes difficult to move a mouse precisely, the usual fault is a build-up of dirt on the rollers (see above) and a thorough cleaning will work wonders. If this does not restore normal operation, check for defective plug contacts and cable connections.

Games ports

A games port is often included as part of an IDE card or a parallel/serial port card. There is no standard for such games ports, and no two seem to be identical in pin-out. The connections to the port refer to buttons and positions for connection to a joystick. Since a PC machine is intended for serious computing uses rather than games, the games port can be disabled.

Chapter **6**

Setting up

Once you have installed all the cards on the motherboard, whether you are working on a machine that you have constructed from scratch or improving an older machine you have bought, you should now turn your attention to the set-up of the whole computer system. This is something that is often neglected, and by spending just a little more time at this stage you can make it all much easier for yourself later. The first point to consider is how you intend to locate all the separate sections that make up a PC. When you are first testing your handiwork, the sections should all be accessible, and the monitor is best placed temporarily on one side of the main casing.

The conventional format is to place the monitor on top of the main casing, with the keyboard in front and the mouse to one side. This places the weight of the monitor on top of the lid of the case, and if the monitor is a heavy one, as all colour monitors are, you should spread the weight with a square of plywood or chipboard placed between monitor and case, so that the edges of the case are taking the weight rather than the more vulnerable lid. This arrangement, though very popular, does not allow you to flip open the lid without first moving the monitor, and a much better arrangement is to place the main casing on a separate shelf, preferably under the desk or table. An old coffee-table of the low variety can be used, and if there is enough clearance above the main case this allows you to flip open the lid without the need to shift anything. This is a much more suitable arrangement if you are likely to be making frequent changes to cards and other aspects of the interior of the machine.

You can place the mouse to the left or to the right, and software will allow you to interchange the functions of the mouse switches to allow for left-hand or right-hand use. Both mouse and keyboard normally come with leads that are long enough to give you considerable choice in where you place them relative to the main casing.

If you have used a maxi-, midi- or mini-tower casing, you are free to separate the monitor from the rest of the machine, and the cables that are supplied with such machines are usually long enough to allow you a lot of freedom in this respect.

Wiring

The standard form of power supply that is used in the PC has a Eurosocket connector that is intended for the power to the monitor or the printer; sometimes both can be connected. This supply may be separately fused inside the power supply unit, or it may share the mains fuse in the 3-pin mains plug of the computer. Some monitors are provided with a Euroconnector, but if none is provided, you can connect your own, Figure 6.1. Euroconnectors are available from the main electronics supply firms such as Maplin and RS Components (ElectroComponents). Remember that the Euroconnector you use for a monitor power lead should be the cable-end pin type to match the socket type used on the computer.

The power cable to the main casing will use another Euroconnector, and you will normally have to fit a mains plug. This must be a standard UK 3-pin plug, and the fuse *must* be a 3A type. Do not on any account fit a 13A fuse, because the internal cabling of the machine is not rated to take such a current without serious damage. If a plug has been supplied, check the fuse rating for yourself, even if you have been assured that it is a 3A type. At this stage, do not insert the mains plug.

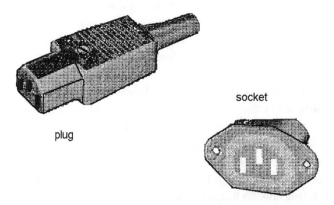

socket

plug

Figure 6.1 *The Euroconnector, shown in plug and in socket form. This is the standard connector for the power supply, in or out, using the socketed version for live leads.*

To connect up the units, you need complete access to the rear of the main case – do not try to insert connectors by feel. The keyboard connector should be inserted first. This is a rather fragile DIN plug, Figure 6.2, and its socket is directly mounted on the motherboard. Locate the key for this plug, and try to use the minimum force when inserting it, because the socket on the motherboard is not particularly rugged. If the plug does not slide easily into the socket, stop and try to find out why – you may be trying to put the plug in with the pins turned to the wrong angle. The keyboard cable is usually coiled, and if it does not stretch far enough in its coiled form, pull it out a bit. You can buy cable extenders if needed.

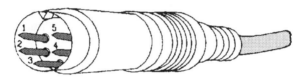

1	Clock pulse
2	Earth
3	Data
4	+5V
5	Not used

Figure 6.2 *The keyboard plug and pin uses – some older Amstrad and modern IBM models differ.*

A keyboard can be easily replaced in the course of an upgrade, though there is seldom any need to do so unless you are working on a very old machine with the 83-key arrangement. Note that the 83-key type and the 102-key type are not freely interchangeable, and the 102-key types often include a switch that can be set for the AT or XT type of machine.

The mouse can now be connected, either to its mouse port if it is the bus type of mouse, or to a serial port. Remember that both serial and parallel ports can use the same form of 25-pin D-connector, but the parallel port uses a female socket. If you have a 9-pin connector on the COM1 port, use this for the mouse. The connectors that are used for this port can normally be screwed into place, but do not screw them down when you are first testing. Drape the mouse cable to one side of the keyboard, leaving enough slack to allow you to move the mouse easily.

Now connect the monitor, inserting the Euroconnector into the PC (female) socket, or plugging in to a mains power point if you have opted to keep the monitor separate from the PC supply. The monitor data plug then has to be inserted. The standard type of 15-pin D-plug will fit only one way round, and even for testing purposes it is advisable to fasten the plug into the socket using the screws at the side. The data cable for a monitor is usually thick, because it uses several sets of twisted leads, and it is also stiff because of metal shielding, so that the connector is likely to be pulled out if you move the monitor unless the plug is fastened in.

Using CMOS-RAM SETUP

The PC keeps some data stored in CMOS RAM, backed up by a small battery that is located on the motherboard. Some motherboards provide for an external battery to be used either together with or in place of this internal one, and if you encounter problems such as a request to alter the CMOS RAM set-up each time you boot the machine, battery failure is the most likely cause. If you have constructed or altered a machine, particularly if you have installed a hard drive, you are likely to get a message when you boot to the effect that an unrecognized hard drive is being used. Along with this you will be asked to press a key to start the CMOS set-up. This type of message is delivered when the machine senses that there is a discrepancy between what is stored in the CMOS RAM and what is physically present, but minor changes such as adding ports will not necessarily affect the CMOS RAM.

If you do not get a CMOS RAM set-up notice when you boot, you may see a notice on the screen notifying you that you can press a key in order to get into the CMOS set-up. The way that is used to make the machine run its SETUP depends on the make of chips that it uses (the chipset). One common method, used with AMI BIOS machines, is to offer you a short interval in which pressing the Del key on the keypad (at the right-hand side of the keyboard) will enter SETUP. The AWARD chipset machines require you to press a set of keys, Ctrl-Alt-Esc, in this interval. Whatever key or key combination is to be used, it should be noted in the documentation for the motherboard, and also on the screen. Note that pressing the *Delete* key (in the set of six above the cursor keys), as distinct from the Del key, will have no effect – this is because the machine is at this stage being controlled by a very small program in the ROM which allows only very limited capabilities.

The snag is that if, as recommended, you have wired your monitor to the Euroconnector so that it is switched on by the computer, the monitor may not have warmed up in time to display the message. Colour monitors in particular tend to miss the message because they warm up slowly. The remedy is to boot up in the usual way, and when the monitor is fully active, press the Ctrl-Alt-Del key combination. Use the Ctrl and Alt keys to the left of the spacebar, and the Del key on the keypad at the right. This key combination causes what is called a 'warm boot', meaning that the computer restarts (clearing its memory on the way), but omits some self-test routines so that the restart is faster. During this restart you should see a message such as:

```
WAIT...
Hit <DEL> if you want to run Setup
```

The AMI BIOS message is illustrated here.

Whichever method is used, it should be possible to see a display such as that in Figure 6.3. This is a simplified example of a modern AMI BIOS chipset display and those for other machines will differ in detail. The important point is that you are offered a set of optional menus to choose from, of which the first (already selected) is by far the most important at this stage. Until you are thoroughly familiar with the system, do not attempt to use any menus other than the Standard CMOS Setup. The only exception is that if you find the system misbehaving after a change in the CMOS set-up you can recover by entering the set-up again and selecting either the BIOS Setup Defaults or the Power-On Defaults. The AMI BIOS set-up reminds you of this when you opt to use either of the main Setup menus.

When you opt for the Standard Setup, you will see a display that is, typically, illustrated in Figure 6.4, and the important point is that the information on the drive types should be present. Any alteration in the installed drives has to be notified, otherwise the CMOS RAM Setup table is likely to be presented to you each time you boot. If you have constructed a machine from scratch, or altered the drives of an older machine, you will certainly need to alter the particulars shown here. Altering a line of information is done by using the arrow keys (cursor keys) to move the cursor to the item(s) you want to change, and pressing either the Page Up or Page Down keys to change the item. Note that you cannot type in numbers or day names, only cycle through the options that are provided. What is less clear is how to find and enter the

```
                            STANDARD CMOS SETUP




                            ADVANCED CMOS SETUP

                           ADVANCED CHIPSET SETUP

                    AUTO CONFIGURATION WITH BIOS DEFAULTS

                   AUTO CONFIGURATION WITH POWER-ON DEFAULTS

                             CHANGE PASSWORD

                            HARD DISK UTILITY

                          WRITE TO CMOS AND EXIT

                        DO NOT WRITE TO CMOS AND EXIT
```

Figure 6.3 *The text of a CMOS RAM display for an AMI BIOS, showing the menu of main options. Some of the hard drive options must not be used nowadays.*

```
Date (mn/date/year)  : Sun, Oct 03 1993           Base Memory : 640 KB
Time (hour/min/sec)  : 10  :  47  :  12           Ext. memory   : 4096 KB
Daylight saving      : Disabled       Cyln  Head  WPcom  LZone  Sect  Size
Hard Disk C: type    : 47 = USER TYPE 1001   15     0      0     17   125 MB
Hard Disk D: type    :  Not Installed
Floppy drive A:      : 1.44 MB, 3½"
Floppy drive B:      : 1.2 MB, 5¼"             Sun  Mon  Tue Wed Thu Fri Sat
Primary Display      :  VGA/PGA/EGA            26   27   28  29  30   1   2
Keyboard             :  Installed               3    4    5   6   7   8   9
                                               10   11   12  13  14  15  16
Month   :  Jan, Feb,...Dec                     17   18   19  20  21  22  23
Date    :  01, 02, 03,.....31                  24   25   26  27  28  29  30
Year    :  11981, 1982,...2099                 31    1    2   3   4   5   6

ESC: Exit   Arrow keys: Select    F2/F3 : Color  PU/PD : Modify
```

Figure 6.4 *Typical standard set-up options of the CMOS RAM. Note that these will not be retained if the backup battery fails.*

information about the hard drive that you need to supply to the system. At this stage in its action, the computer cannot make use of the mouse, and only a few keys, such as the cursor and Esc keys, are recognized.

Moving the cursor to this line, the *Hard Disk C:* line, allows you to use the Page Up and Page Down keys to alter the disk type number from 01 to 47, with 47 being used as a way of entering information that cannot be covered by any of the others. Many types of IDE drives allow you to use a standard number such as 17, even though the drive is quite differently organized. This is because the circuits on the drive can create an emulation of the standard type. This is why an IDE drive must *never* be low-level formatted, because a low-level format action could possibly destroy this emulation. The documentation that comes with the drive should state which number to use, or, failing that, what figures to use for a Type 47 entry.

If your motherboard is a 486 or later type, the BIOS may be one that will recognize a hard drive automatically, so that though a CMOS RAM panel exists, it is not needed when you install a hard drive. The most recent BIOS system allows Plug'n'play for all devices, cutting out all setting up or use of jumpers, but this is fully usable only if three conditions are fulfilled:

- The BIOS must include the Plug'n'play codes.

- The device you are fitting must be to Plug'n'play standards (a sticker on the label will show this).

- You are using Windows 95 or a later version.

Hard facts

This system of numbered drive types arose in the early days of the IBM PC/AT design (in 1981), when a hard drive of 30 Mbyte was considered as large, and the table contains drive types that are most unlikely to be encountered nowadays. As the system does not cater for really large drives, it is fortunate that the Type 47 is available, and equally fortunate that IDE types can generally use a fictitious number.

This is not helpful if you have just acquired an IDE drive of unknown specification from a car-boot sale (should you have such a trusting nature?) and you want to use it. Laying aside the matter of buying technical products in such a way, not to mention the question mark that must hang over their origin if the seller cannot tell you anything about the drive, you can generally set up the drive if you know its capacity. By cycling through the types in the list you are likely to find one that looks right in terms of capacity, and you can try this. You could also try Type 47 and select figures that lead to the promised capacity. It's all hit and miss, and worthwhile only if the drive was very inexpensive. Incidentally, a drive bought in this way which is not an IDE type is unlikely to be a bargain at any price – it may be an old MFM or RLL drive, in which case your IDE card cannot interface to it and the drive is likely to be an old one whose MOT is imminent (and it will probably fail).

Floppy and display details

Once the hard drive details have been entered you must fill in the portion of the Setup form that deals with the floppy drives. Place the cursor on the line for *Floppy drive A* and use the Page Up and Page Down keys until you see the type of drive you have installed, usually a 3½" 1.44 Mbyte type – note that this counts 1 Mbyte = 1000 Kbyte rather than 1024 Kbyte. If you have a second floppy drive, declare the details of this also, again selecting by using the Page keys. There is a *Not Installed* option which is used for machines that are part of a network and which do not need disk drives – this might be the selected option if you have obtained a motherboard from a networked machine or you are refurbishing such a machine.

The display line is selected in the same ways, with the options of Not Installed, Monochrome (meaning Hercules), Colour 40 x 25 (meaning CGA), VGA/PGA/EGA, and Colour 80 x 25 (no graphics). The Not Installed option would be used for server machines on a network, and, as before, if you are re-furbishing such a machine or using its motherboard you might find this option set. The normal option is VGA/PGA/EGA, whether you use a colour or a monochrome VGA monitor – the meaning of PGA is lost in the mists of time. The only keyboard options are Installed or Not Installed; once again, a network server might be run without a keyboard. The *Daylight Saving* option is an odd one which few, if any, motherboards support, and it should be ignored.

You can then correct the calendar and clock details if necessary. The timekeeping of the PC machine's clock is notoriously poor, and you may find the time incorrect by more than a few minutes (sometimes by more than an hour). This is not a battery fault, but one that is caused by the interrupt system which momentarily stops the clock when other actions are needed, so that the clock is always slow. The calendar details are usually correct unless the board was not set up correctly initially, or the battery has failed. You are not obliged to set the calendar and clock at this stage, but it is useful to do so. If you want to correct the time later it can be done using the MS-DOS TIME command rather than by using the CMOS RAM option.

Booting up

You can now leave the CMOS set-up program, taking the option to *Write to CMOS and Exit*, and when you boot the machine the hard drive should now be recognized. If you are using a hard drive that has been in service, it may be provided with MS-DOS. If this is so, you should *see* the message:

```
Loading MS-DOS.....
```

(or something similar) appear, and hear the disk working hard. You may find other items of software on the disk. Sometimes a new hard drive comes ready for use in this way, though it is more normal to require formatting. If instead of the Loading MS-DOS message you get a message such as:

```
Error loading operating system
```

or:

```
Non-system disk or disk error
```

then the hard drive is almost certainly not formatted. You will need to insert an MS-DOS system disk in Drive A and reboot (press Ctrl-Alt-Del) to load in the operating system. This process is described in detail in Chapter 8 and will not be repeated here.

Advanced CMOS Setup

You would not normally use the CMOS Setup advanced section when you were first booting up a new or refurbished machine, but for future reference it is useful to know what this part of the set-up can accomplish. The difficulty here is that there is no standard set of actions, and the examples are taken from two samples of AMI BIOS chips – BIOS chips from other manufacturers, such as Award, are likely to differ, though several options will be the same. Some of these options can be used without the need for deeper understanding of the computer, but those which deal with memory allocations, particularly with ROM shadowing (see later) should be left strictly alone until you know what is involved on your machine. A typical list is shown in Figure 6.5.

```
Typematic Rate Programming        :  Disabled
Typematic Rate Delay (ms)         :  250
Typematic Rate (Char/sec)         :  30.0
Above 1MB Memory Test             :  Disabled
Memory Test Tick Sound            :  Disabled
Memory Parity Error Check         :  Disabled
Hit <ESC> Message Display         :  Disabled
Hard Disk Type 47 Data Area       :  0:300
Wait For <F1> If Any Error        :  Disabled
System Boot Up NUM Lock           :  On
Numeric Processor                 :  Absent
Floppy Drive Seek at Boot         :  Disabled
System Boot Up Sequence           :  C:,A:
Internal Cache Memory             :  Disabled
Password Checking Option          :  Disabled
Video ROM Shadow C000,32K         :  Disabled
Adapter ROM Shadow C800,32K       :  Disabled
Adapter ROM Shadow D000,32K       :  Disabled
Adapter ROM Shadow D800,32K       :  Disabled
Adapter ROM Shadow E000,32K       :  Disabled
Adapter ROM Shadow E800,32K       :  Disabled
System ROM Shadow F000,64K        :  Disabled
Memory Wait States                :  Disabled
RAS Time Out                      :  Enabled
16bit ISA Cycle Insert Wait       :  0 ms
RAS Active Timer Insert Wait      :  Disabled
Quick RAS Precharge Time          :  Disabled
Slow Refresh                      :  Enabled
IO Recover Period Define          :  Disabled
```

Figure 6.5 *A typical set of advanced CMOS options.*

One set of advanced options deals with the Typematic rate. This is the rate at which a key action will repeat when a key is held down, and there are two factors, the time delay between pressing a key and starting the repeat action, and the rate at which the key action repeats once it has started. The set-up options are to enable or disable the Typematic action, to set the delay, and to set the repeat rate. Typical default values are to have the action disabled, the delay set to a long 500 milliseconds (0.5 second) and the rate to a fast 150 characters per second. My own settings are Enabled, 250 milliseconds and 20 characters per second. This is all very much a matter of choice, and you need to remember that some software, particularly Windows, will override these settings and impose its own.

Another group of set-up options deals with memory testing, and though this set might look intimidating there is, in fact, no harm in altering the settings because they are no longer of prime importance in a modern PC. These options are concerned with the automatic testing of memory by the PC when it is booted up. The original PC machines could use a maximum of 640 Kbyte of memory, so that originally only this amount of memory was tested, using another set of memory chips to hold parity bits (used for checking for any change in a byte). In 1979, the type of memory chip (Dynamic RAM, or DRAM) that is so common nowadays was regarded with great suspicion, and it was thought essential to test the memory each time the machine was switched on. In fact, memory is now quite remarkably reliable, and the checking of memory is, for most purposes, a waste of time. The checking needs to be on only if memory is suspect, and a few machines install memory in 8-bit units only, omitting the ninth bit that was traditionally used for checking.

The three options are Above 1 MB Memory Test, Memory Test Tick Sound, and Memory Parity Error Check. The Above 1 MB Memory Test can be disabled to cut down the time waiting for the machine to boot, because on modern machines with 2 Mbyte or more of memory this takes quite a significant time and is needed only if you suspect memory problems (which are so rare as to be most unlikely). Oddly enough, MS-DOS 6.2 now incorporates a check of extended memory in its HIMEM.SYS controller (see later).

The ticking sound that accompanies the memory check can also be disabled if, like me, you prefer your computer not to be incessantly hooting at you. The Memory Parity Error Check can likewise be disabled to cut out all memory checking (including the first 1 Mbyte), on the same basis that memory is reliable and needs none of this. This latter option must always be used if the memory is made up from 8-bit SIMMs.

There are two options that are concerned with the set-up itself. You can opt to enable or disable the message that tells you which key to press during booting in order to run the set-up. This is a small precaution against anyone re-setting your CMOS settings by accident or design. The other option is to enable or disable a message about pressing the F1 key if an error occurs during booting. Disabling this option will allow the CMOS set-up to be displayed at once if such an error occurs.

There are usually several options that allow for easier and faster booting and subsequent use. The System Boot Up Num Lock option should be enabled, so that the number-keypad on the right-hand side of the keyboard is set for numbers rather than its optional cursor keys. This avoids the need to have to press the Num Lock key after booting. The Numeric Processor Test should always be disabled unless you actually have a numeric co-processor installed (which is rare) – even if this chip is installed it does not really require testing unless you suspect that it might be faulty.

A really useful option is to disable the Floppy Drive Seek at Boot Time. Normally, the machine checks both floppy drives and will attempt to boot from drive A if there is a disk in this drive. You will want to boot the PC from its hard drive, except for the first time before you have the hard drive formatted and loaded with the operating system. Once this has been done, the automatic attempt to boot from the floppy drive is a waste of time, particularly if you have left a disk (other than a system disk) in Drive A. The effect of a non-system disk in Drive A is that the machine will try to boot from this disk and provide you with an error message when it finds that the MS-DOS tracks are not on this disk. Quite apart from the damage to your blood pressure as you wonder if the hard drive has disintegrated, this is time-wasting because you need to remove the floppy and re-boot. By disabling the Floppy Drive Seek, the machine will not automatically try to check the floppy drives.

The other option that is associated with the floppy drives is the Drive Seek Sequence. The default is A:,C:, which tries to boot from the floppy drive first. If this is altered to C:,A:, the machine will always boot from the hard drive unless there is some form of hard drive fault (and even such faults are often blips and do not recur after you have re-booted). This allows you to keep a floppy permanently in this drive to use for backup.

If you have an option to set the CPU speed to its fastest rate and to disable the Turbo switch you should take advantage of both options, because both are great time-savers. With a Turbo switch on the front panel it is just too easy to switch to a lower CPU clock speed by mistake, and not to find the error for a considerable time.

The no-go areas

There are several advanced settings options that you should avoid for the present, and possibly for ever. There is usually an option to place information on a Type 47 hard drive in one of two areas of memory. The conventional place is noted as 0:300, and this should be used unless you know something to the contrary. The alternative is described as DOS

1 KB, meaning 1 Kbyte of the memory that is used by MS-DOS. Avoid this unless you are certain that the normal area of memory is being used by something else and that the alternative is safe. For most of us, this means leave it alone.

The main no-go areas for the present are concerned with ROM Shadow. The ROM type of memory is permanent but slow to read, and most PCs can gain a significant speed advantage by copying the contents of ROM into fast RAM during the boot time. Before you whoop with delight and switch this facility on you need to make certain that the memory that will be used in this way will not also be used by something else. If you have 2 Mbyte of memory and are using DOS only (not Windows), take the option to shadow Video ROM and System ROM by all means and reap the benefit. If you are using Windows and MS-DOS 6.0, and intend to employ the utility MEMMAKER to set up the use of the memory, be very careful because MEMMAKER, along with other utilities that optimize the use of memory, does not necessarily recognize that a piece of RAM is being used for ROM shadow. Do not use the Adapter ROM Shadow options unless you know that you have a card that contains ROM chips in the address ranges that are shown. For most users, the System ROM and Video ROM are the only two that should be shadowed.

Finally, the Password options allow you to create a password either for access to Setup only, or for access to the whole system. Passwording can be useful when a machine is available to a large number of people, but unless you have security problems it is best to avoid passwording. For one thing, you need to remember your own password. If a password is easy to remember, it is usually easy for someone else to guess. If you forget a password you will be locked out of your own machine and there is no simple way then of disabling the passwording, though it can be done by an expert. If you are desperate, the AMI BIOS provides for the password changing to AMI when the back-up battery is discharged or momentarily disconnected.

The Hard Disk Utilities

The CMOS main menu usually contains a Hard Disk Utility option. This is a hangover from days past, and it contains two utilities that should not under any circumstances be used on a modern IDE or SCSI drive. These are Hard Disk Format and Auto Interleave. The Hard Disk Format is the low-level type, not to be confused with the high-level format which is needed for a new hard drive of any variety. The low-level format was needed for the older types of hard drives, known as MFM, RLL or EDSI, and it ensured that the tracks and sectors were prepared for use. Modern IDE drives have been low-level formatted during manufacture, and they are usually set to emulate some standard number. Any attempt to perform a low-level format is likely to upset this arrangement and result in a disk which will, at best, not store to its full capacity.

Interleave is also a relic of the past when disk controllers and RAM could not keep up with the rate of flow of data to or from a hard drive. Using a 1:2 interleave, for example, meant that the disk would read or write one sector on a track, and then spin around one more time before dealing with the next sector, rather than dealing with the sectors in order during one revolution of the disk. Interleave factors as high as 1:6 were used on some machines, but modern IDE and SCSI drives have a 1:1 interleave which must not be changed.

The only utility that might be useful is the Media Analysis which will check each track on the hard drive, and mark off bad tracks so that they are not used. This is not usually needed on modern drives, and there is software which can provide more extensive reports than are available from the Media Analysis and which should be used in preference. If you use MS-DOS 6.0 with DoubleSpace (see the *MS-DOS 6.0 Pocket Book*) you should avoid all utilities that might alter the distribution of data on the disk.

Chapter **7**

Upgrading to 386

This chapter is intended particularly for the reader who has taken the simple and cost-effective route to a modern PC – upgrading an old machine. The machine may be one that you have had for some time, an old 80286 machine, for example, or it may be one that has been bought at a very low price at auction or in a car-boot sale (where machines recovered from skips outside hospitals or offices often end up). Another route for the upgrader is to buy a new machine which is at a rock-bottom price because it contains the essentials of a case, power-supply and motherboard only. There are fewer suppliers of such machines nowadays, but the prices for such machines are lower than you would have to pay if you bought the parts and assembled them for yourself.

Machines bought at auctions, particularly from bankrupt office firms, can present you with an interesting gamble. Some of these are likely to be very powerful machines, particularly if a network has been in use and you can lay your hands on the server machine. You can also find fast and efficient networked machines with no disk drives, but with a fast processor and a lot of memory. You have to be able to look inside the cases for yourself and determine, from the processor type-number and the number of SIMMs you can see, whether a machine is likely to be a bargain or not. Do not depend on an auctioneer's description, because they are not computer experts and have to rely in turn on whatever description they can find. Goods bought at auction are usually bought 'as seen' with no form of warranty as to content or origin.

In general, if a machine has the 80486 chip on its motherboard and 8 Mbyte or more of memory, it will be snapped up by a dealer who can refurbish it and resell, so that you are not likely to get the chance to buy it cheaply for yourself. That said, a small local auction may not attract any dealers who know what to look for, and you can be very fortunate. The usual bargains at the larger auctions are the 80386DX machines, and if you can find one with a decent amount of memory and a hard drive, then it's no hardship to part with £100, which is often as little as you need to secure it now that the 80386 is unfashionable with the jet set. Think carefully about 80386SX machines unless they have a lot of memory, because the 80386SX motherboards are now quite cheap. Do not consider 80286 machines, unless at very low prices, because to bring them up to date you will need a new motherboard and probably new drives as well, and avoid 8088 or 8086 machines which require new *everything* to bring them up to date (and some of them are almost impossible to upgrade).

Even some modern machines, usually bearing famous names, are difficult to upgrade, which is why they found their way to the auction in the first place. Read all of this chapter before you consider this path to a new PC, because trying to upgrade some machines could be more expensive than starting from scratch. Avoid in particular machines with plastic, or very slim metal, cases. Avoid any whose drives are stamped with the same name as is on the case – a standard drive might not fit in the same space; if the name on the case is unfamiliar you are more likely to be in luck. Machines which use unfamiliar layouts, 72-pin SIMMs, and disk interfaces on the motherboard should all be avoided – they are likely to be excellent machines, but upgrading them will be costly and difficult.

Drive change

One common drive upgrade route starts with the machine that has no hard drive, and sometimes no floppy drive, having been used as a workstation on a network. Another is the machine which, though provided with an 80386 processor and adequate memory, has an old hard drive of the MFM type and small capacity, such as 20 – 40 Mbyte. Such a machine might use an old 5¼" floppy drive as well. Another common situation is a machine that uses a 720 Kbyte floppy drive rather than the modern 1.44 Mbyte type. All of these examples call for refurbishment in the drive bay department.

The comments in Chapter 3 apply to drive installation in general, and what follows is aimed particularly at the user who is stripping out an old drive or adding a drive, rather than installing a new drive into a new case. Obviously, the methods of installing and using a drive are the same. What you need to look out for are machines of eccentric design and with unorthodox fastenings. Nameless clones are a delight to work on, because they follow a well-worn standard pattern. Big-name machines can be a nightmare, because only their own spare parts will fit, and these parts are usually three times as expensive as the (almost identical) part for the nameless clone.

The most pressing need for a drive occurs when the machine is a diskless workstation, or the workstation machine has only a floppy drive. These machines have been networked to a very fast and larger server machine which provided the hard drive capacity and printer port for the separate workstations, but you often find that each workstation has a fast processor and plenty memory. The only thing to watch out for is that the casing is standard, providing three or more drive bays. Some workstations use an ultra-slimline casing which has no space for drives of *any* description, and this would not be a good buy even at a very low price, unless you fancy your ability in the tin-bashing department.

On a diskless machine, check that the power supply unit has connectors for disk drives. These are usually provided on clones because the power units are of a standard design, but if the supply has no connectors for drives you cannot proceed unless you buy a new power supply or add a set of drive cables to the existing one. In the absence of a circuit diagram and layout diagram for the power supply unit this would not be advisable, but if you are familiar with the type of switch-mode supply that is used you can make up your own cables – always provide at least four drive connectors so as to allow for future expansion.

Now check the drive bays. You may need to loosen off the front panel to gain access to the lowest bay so that you can secure a drive to its sides – the bolt holes are usually inaccessible when the front panel is in place and the drive bays fastened down. Make sure that you have fastening bolts of the correct type for your drive(s). If the drive bay container is of the type that allows a 3½" hard drive to be mounted sideways you are in luck, because this makes the installation of the hard drive much easier. If the sideways form of drive fastening is not included, you will need an adapter plate for a 3½" drive. You will not normally be using a 5¼" hard drive unless you have obtained a large-capacity unit of this size at a bargain price. This is not particularly likely, and the large units are often intended for use with a SCSI connection rather than an IDE one. SCSI connections on PC machines are not so thoroughly standardized (or so common) as the IDE type, so you are on your own unless you can obtain manuals for the drive and its interface card.

Always install the hard drive first, in the lowest bay or in a sideways position if this is available. If no sideways fasteners are provided, you will need to bolt the drive into the lowest bay and then bolt the set of bays into the case – after checking that the drive is firmly in place. Replace the front panel if you have had to remove it. Connect the power cable connector to the power connector on the drive, making sure that the two are correctly engaged. The point about using the lowest bay is that this one has a permanent front cover – the bays above will have removable front covers so that floppy drives can be installed.

If you also need to install a floppy drive, do so now, or replace whatever floppy drive was on the machine if it lacked only the hard drive. Remember that a 3½" drive will often require an adapter for its power supply, and these are not always easy to find in local shops. If the machine has come with a 3½" floppy drive included, check that this is not a 720 Kbyte drive – you will need the 1.44 Mbyte variety for most software packages. The cost of a floppy drive of either type is around £30 at the time of writing.

You may want to add a second floppy drive. It may be useful to be able to work with 5¼" disks as well as the 3½" type, and a drive of this size is worth considering. On the other hand, you might prefer to be able to work with two 3½" drives, allowing for easy backup of software on this type of disk.

As a small note on the side, the main difference between a 3½" 1.44 Mbyte floppy and a 3½" 720 Kbyte floppy is the second hole in the casing. Some modern drives allow you to treat a 720 Kbyte drive as if it were a 1.44 Mbyte type, ignoring the absence of the hole; older drives in general will use the presence of the hole to indicate a high-density disk. This hole need not be square, and a hole of around ⅜" diameter in the right place will work wonders, Figure 7.1. Of some two hundred 720 Kbyte disks I have drilled in this way, only one refused to format as a 1.44 Mbyte disk – and that was a disk given away with a magazine. At the present price difference, fifteen minutes with your Black and Decker is time well spent – but make sure that all the dust and swarf is totally removed before you use the disks.

At the upgrading stage, you might want to consider drives other than straightforward magnetic drives. The options currently are floptical disks, tape streamers, and read/write optical drives. These latter drives are still extremely expensive, and unless you were able to find one on a bargain machine they are not currently a popular item for building in to a DIY machine. Floptical drives are conventional magnetic drives which use a laser guidance system to write much narrower and close-packed tracks. This allows a 3½" disk to hold some 21 Mbyte, a useful size for backing

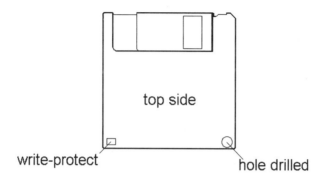

top side

write-protect

hole drilled

Figure 7.1 *Drilling a hole in a 720 Kbyte 3½" disk to convert it to 1.44 Mbyte. This is frowned on but is widely practised.*

up files. The drives are not cheap compared to conventional 3½" types, though, and for backup purposes (meaning that you will not necessarily need to read the files back), a tape streamer can be much more cost-effective.

Currently, tape streamers such as the Colorado Jumbo in the 60 and 120 Mbyte sizes are inexpensive, and the tape cartridges that they use are also inexpensive. If you need the peace of mind that backups provide (or you feel that your bargain hard drive might not be such a bargain for long) then a streamer of this type can be a good buy, and is installed in much the same way as a conventional 5¼" floppy drive.

Remember that when you have made changes to floppy or hard drives you will have to notify these changes to the CMOS RAM of the machine, as detailed in Chapter 6.

Graphics card

If you are upgrading, the need to upgrade the graphics card is less likely than the need to upgrade disk drives, because the VGA type of card has been a standard for some considerable time. Machines that used a CGA card are likely to be thoroughly obsolete by now, and hardly worth the trouble and expense of upgrading. You might, however, want to upgrade the graphics card so as to speed up the use of Windows or to be able to use SVGA modes such as 800 x 600 graphics.

Your choice of graphics card is important if you want to perform such an upgrade. Upgrading from CGA or Hercules calls for little more than a conventional VGA card, and there are plenty to choose from in the £30 cost region, but going for fast action and/or SVGA needs more consideration, as Chapter 4 has explained. If you are also upgrading a motherboard and going for an 80486 type, the chances are that the new motherboard will use some local bus slots, which operate at a much higher speed than the ordinary slots. If you want to take advantage of this extra speed, you will need to buy a graphics card that is designed for use with a local bus, and you also need to be certain that it is compatible with the local bus on your motherboard. You can expect to pay up to £300 for such a graphics card (though the lower-cost cards are very effective, and faster than almost any card on a conventional bus); this applies also to some of the ultra-fast cards that use the ordinary bus structure and which can be used on 80386 machines.

Ports

Upgrading ports is simple enough in mechanical terms, the insertion of a card, but you need to give thought to setting jumpers or DIP switches before inserting the card. The main problem is that if, as is usual, you have one parallel port and at least one serial port on the IDE card, any port card you add is likely to cause some conflict with the existing ports unless you are careful to select the correct address and interrupt settings. Chapter 5 contains details of what to look for. One important reason for upgrading ports is likely to be the use of a modem, and since this is a topic that requires some thought and a lot of information, much of this chapter will be given over to it.

Modems and serial communications

The modem is a device that makes use of a serial port to transmit or receive data one bit at a time. When data is sent one bit at a time, some method has to be used to allow the receiving computer (or other device) to distinguish one group of eight bits (a byte) from its neighbour. This problem does not exist for a parallel system, because in the instant when a character is transmitted, all of its bits exist together as signals on the eight separate lines. For a serial transmission there is just one line for the bits of data, and the eight data bits must be sent in turn and assembled into a byte by storing them at the receiving end. The problem is that since one bit looks like any other, how does the receiving machine recognize the first bit of a byte?

The way round the problem is to precede each transmitted byte of eight bits by a start bit (a zero) and end it by either one or two stop bits (each a 1) – notice that once again there is no standardization of the number of stop bits, though one stop bit is slowly becoming the more common practice. Ten (or eleven) bits must therefore be transmitted for each byte of data, and both transmitter and receiver must use the same number of stop bits. The transmitting computer will send out its bytes of data; at the receiving computer, the arrival of a start bit will start the machine counting in the bits of data, storing them into its memory until it has a set of eight, and then checking that it gets the correct number of stop bits after the last data bit. If the pattern of a zero, then eight bits (0 or 1), then a 1, is not found (assuming eight data bits and one stop bit), then the receiving computer can be programmed to register a mistake in the data, and start counting again, looking back at the stored data and starting with the next 0 bit that could be a start bit. The recounting is fast, and can be carried out in the time between the arrival of one bit and the next, so that it would be unusual to miss more than one character in this way. All of this is the task of the communications software that you use with each computer and is carried out automatically if, and only if, you have programmed the software to work with the correct settings.

The use of the same number of stop bits and data bits by both computers is not in itself enough to ensure correct transfer, though. In addition to using the same number of data bits and stop bits, both transmitter and receiver must work with the same number of bits per second. In the (rather loose) language of computer communications, they must use the same baud rate. Baud rate and bits per second are not necessarily identical, but that's something for the experts to worry about. Quoted figures for the speed of communications always use the phrase baud rate to mean the number of bits per second, and we'll stick with that to avoid confusion. Figure 7.2 shows the RS-232 standard baud rates, of which 300, 1200, 2400 and 9600 are the most common nowadays.

The rates below 300 baud are hardly used other than by the painfully slow old-style Prestel rate of 75 baud (for transmitting), and even 300 baud is becoming less usual now that better telephone links and better software are available. For most practical purposes you can take the baud rate as meaning the number of bits per second, so that the number of

50	75	110	150	Slow, seldom used
300	600	1200	2400	Printers & modems
4800	9600			Modems and Fax
19200				Terminals

Figure 7.2 *RS-232 Baud rates. Rates of up to 19200 are supported by MS-DOS, but the hardware can handle higher rates of more than 100000 bits per second, given suitable software, such as Interlink of MS-DOS 6.0.*

characters per second is equal to the baud rate divided by the total number of bits in each byte. If, for example, you use a start bit, eight data bits and one stop bit, you have a total of 10 bits per byte, and at 300 baud you will be sending 30 characters per second. It's certainly faster than you can type, but at this speed you will need a time of around 800 seconds (13.3 minutes) to transmit a 4000 word article, and the timer will be ticking away inside the telephone exchange for all this time, plus the time it has taken you to make sure that transmission has been established.

All serially transmitted data will almost certainly use ASCII code and if you are transmitting ASCII text only (not program data) this will require only 7 of the 8 data bits that can be sent. If only 7-bit ASCII is needed, then the eighth bit can be used as a parity bit, a check on the integrity of the data. The parity system can be of two types, even or odd. In the even parity system, the number of ON signals (logic 1s) in the remainder of the byte is counted, and the parity bit is made either 1 or 0 so that the total number of 1s is then even. In the odd parity system, the parity bit will be adjusted so as to make the number of 1s an odd number. Figure 7.3 shows what this involves – it's all done inside the computer, and the only effort on your part is to specify whether it should be done or not, and, if so, which parity to use. At the receiver, the parity can be checked and an error reported if the parity is found to be incorrect. This simple scheme will detect a single-bit error in a byte, but cannot detect multiple errors nor can it correct errors. Since it is applicable only when 7-bit data is being transmitted, it is used mainly for text transmissions, and nowadays is often omitted altogether. This is, incidentally, the same as the method used for checking memory by using a parity bit.

Good software can make use of other methods of adding data so as to provide for better checking and can even provide for the correction of

```
1 1 0 1 0 0 1    seven-bit signal
0 1 1 0 1 0 0 1  add even-parity bit
1 1 1 0 1 0 0 1  add odd-parity bit
0 1 1 1 1 0 0 1  error, even-parity
1 1 1 0 1 1 0 1  error, odd-parity
```

Figure 7.3 *Parity illustrated by showing a 7-bit signal and how the eigth is added according to the parity. A single error will make the parity bit incorrect (the number of 1 bits will be odd instead of even or vice versa).*

errors to some extent. The checking methods can range from the simple checksum to the very complicated Reed-Solomon system (also used in compact discs), but they all have one factor in common: redundancy. All checking involves sending more bits or bytes than the bytes of the data, with the extra bits or bytes carrying checking and error-correcting signals. Some of these can work on individual bytes, even on individual bits; others are intended to work on complete blocks of 128 bytes or more. The checksum, for example, works by adding the number values of all of the ASCII codes in a set number of bytes, often 128. This sum is transmitted as a separate byte, and at the receiver the codes are again summed and the total compared with the transmitted checksum. Only if the two match will the set of bytes be accepted.

Because these checking methods all involve the transmission of extra bytes, they slow down the rate of communication of useful data. For text transmissions, the use of elaborate checking is often unnecessary because an occasional mistyped character in text is often not important compared to the need for a high speed of transmission. For sending program files, however, one false byte is usually enough to ensure that the program does not run correctly, so that much better checking methods must be used even if this means taking longer to transmit the data. You often have to select, by way of your communications software, different methods for transmitting different types of data. If you use communications for purely business purposes you are likely to be concerned mainly with exchanging text rather than program files, but you might need to use exchange methods that employ checking if you transmit, for example, your Lotus 1-2-3 data files from one office to another.

Protocols

The individual items of number of data bits, number of stop bits and the use of even, odd or no parity, make up what we call the modem serial *protocols*. You can't get very far in communications without knowing something about protocols, because unless both the transmitter and the receiver are using identical protocols there will be no communication, and only gibberish will be received. There is no single protocol that is used by everyone, so you need to be able to set your communications software to the protocol that is being used by the machine to which you want to be linked. Modern software makes this considerably easier than it used to be, but you still need to know what protocols are being used by the computer with which you are trying to communicate. Communications is just about the last part of computing in which you cannot bridge a gap in technical knowledge with clever software, though as we shall see, it is possible to use software that requires only an initial effort from you.

With suitable software you might set up a few disk files of protocols so that when you wanted to contact a remote computer you only had to specify one of the patterns of protocols that you had recorded. At the most, you would only have to type answers to on-screen questions about data bits, stop bits, parity and handshaking, and all this would be done *before* you started to clock up time on the telephone. On the other hand, this type of software often requires you to be able to write a set of instructions in a formal way, rather like using a computer programming language. If you *have* used a programming language such as BASIC, then you will find this requirement simple enough, but if you have no experience at all of programming, then the task can be rather more daunting. This book can give only the broadest guide to such programming, because each piece of software uses its own methods.

Given that the transmitter and the receiver are set up to the correct protocols, meaning that the baud rate, the number of stop bits and the use of parity will be identical, we still need some method of handshaking to ensure that signals are transferred only when both transmitter and receiver are ready. You might, for example, be recording data on your disk as it arrived, so that there would have to be a pause in the reception of signals while the disk system got into action. Another possibility is that you are printing out the data as it is transmitted, and your printer operates at a speed slower than the baud rate of the transmission. Whatever the reason, just to have transmitter and receiver working at the same rate is not enough, because you also have to ensure that the bits are in step at all times, and you have to make it possible to pause now and again without any loss of data. A serial link to a printer can make use of 'hardware handshaking' meaning that the handshaking can be implemented by using electrical signals over another set of lines, but this option is not open to most communications applications because we can't use extra telephone lines.

The handshaking is therefore implemented in software by using the XON/XOFF system. This uses the ASCII code numbers 17 and 19 (which are not used for characters) sent between one computer and the other. Data can be sent out from a computer following the ASCII 17 code, and disabled following the ASCII 19. Since these codes are sent over the normal data lines, no additional electrical connections are needed. The rate of data transfer is slower because of the time that is needed to send the XON/XOFF signals, so that if you organize your system in such a way that the least use of handshaking occurs, you will transmit or receive faster. The use of XON/XOFF is by far the most common method of software handshaking that is used in communications (another system is called EIA).

Computer to computer

Suppose we have two computers, which could be of any two types, in the same room or the same building and we want to transfer data between them, despite the fact that they use different disk formats. If each computer has a serial port, the input/output connector for serial signals, then all that is needed is a cable to join these ports. If the cable can be bought or made to order (it would have to be of the non-modem or DTE-to-DTE type) then comparatively simple software can be used to transfer data that is entirely composed of ASCII codes, and almost any kind of communications software can be used to make the task considerably easier – the INTERLINK of MS-DOS 6.0 allows a simple serial link like this to be used as a form of network. The methods that we use for more remote computers all centre on the use of a modem.

Consider for a moment the achievement of Alexander Graham Bell in 1876, which he beneficently intended as an aid for the deaf, but which we now call the telephone. The telephone system, now as then, is intended to transmit the electrical signals that are obtained from a microphone operated by the human voice. Like any other waves, the waves of sound of the voice cover a range of frequencies, meaning the number of vibrations per second. In honour of the pioneer of radio, Heinrich Hertz, the unit of one vibration per second is called the hertz, abbreviated to Hz. The range of frequencies that is needed for intelligible (as distinct from high-quality) speech transmission is quite small, of the order of 300 to 3000 vibrations per second, usually written as 300 – 3000 Hz. By contrast, the transmission of music of the quality that we get on a good recording demands (but seldom gets) a range of frequencies from about 30 Hz to 18000 Hz, and television pictures require a range from zero hertz to 5.5 million hertz because of their complicated pattern. The equally spiky signals from a computer also, even if they are slowed down to a low speed of 300 per second or lower, simply cannot be transmitted through a telephone circuit. What we need is a device that detects a computer ON signal and turns it into one tone and will similarly turn an OFF signal into another tone, or a tone that is different from the first in some way, like being out of step. The action of converting the computer signals into tones is called modulation, and we must be able also to carry out the opposite action which is called demodulation. The combining of the words modulation and demodulation gives us the word modem, which is the name of the device that carries out this transforma-tion. The modem is connected to the computer by a serial link, using normal hardware handshaking, but the lead out from the modem to the outside world is a single lead, suitable for telephone connection, using XON/XOFF handshaking.

For most purposes, you need to use a modem that will work in both directions at the same time (full-duplex) so that just altering one tone is not enough, and you need to be able to work with two. For low baud rates, it's possible to have four different tones in use, two for transmitting and two for receiving, with the actual frequency of the tones carefully chosen so as not to cause problems with the telephone system itself. Most telephone systems around the world use different tones to represent the different dialled numbers, and even in the UK where the old pulse system (one pulse for one, two pulses for two and so on) is still used for domestic telephones, the tone system is used between exchanges.

For the higher baud rates the use of separate tones is not possible. You cannot carry a signal of 2000 baud with a frequency that is lower than 2000 Hz, and because the higher baud rates are much too close to the upper limit of frequency that the telephone lines can cope with there is

master (clock) wave

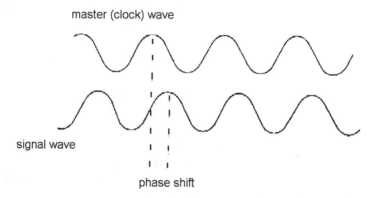

signal wave

phase shift

Figure 7.4 *Illustrating phase shift: the signal wave is moved to lead or lag the standard (clock) signal and the amount of movement is interpreted as a bit or a set of bits. For example, with two phases, one can represent 1 and the other zero. With four phase shifts, you can represent the two-bit combinations of 00, 01, 10 and 11. This allows more bits per second to be transmitted without too much alteration of the signal frequency.*

no possibility of using four tones each higher than the baud rate and each separable from the other. The faster modems therefore use a system of phase shifting, meaning that the frequency is unchanged but a change is represented by a shift (called a phase shift) of the signal, as illustrated in Figure 7.4. This allows a signal of comparatively low frequency to carry several thousand changes per second, so that a rate of 2000 bits per second can be carried with a signal that need be sampled only 600 times per second. This is where baud rate and bits per second become different, and the established practice as far as computer communications are concerned is to refer to the rate in bits per second as the baud rate. As a computer user, the number of bits per second is the important figure for you, the true baud rate is the figure of importance to the telephone line engineer. The more modern telephone exchanges which are being installed all over the country will cope with these phase shift signals, but where the old fashioned 1910 style (that's the date of invention, not the type number) of automatic exchange is still in use, you can use only the slower baud rates.

Now at this point it's important to clear up yet another misuse of words that creates a lot of problems for users. Practically all modems that are used for computer communications are full-duplex, using one set of frequencies (or frequency change) to transmit and another to receive. A full-duplex *modem* attached to one computer must be connected through the telephone lines to another full-duplex modem at the other end, because if it were not then the frequencies or frequency changes would not match. If you had a full-duplex modem connected to a half-duplex modem (using the same frequencies for transmitting and for receiving) then it could receive the signals from the full-duplex modem but could not transmit to it.

The confusion arises because the terms full-duplex and half-duplex are applied also to software that controls the use of the modem. A lot of services that you contact with your full-duplex modem will echo back your signal as you send it so that a copy appears on your screen. Software should refer to this as remote echo, but sometimes calls it full-duplex. Other services do not carry out this remote echo, and your software may instruct you to switch to half-duplex, meaning local echo (so that what you type is put on to your screen by your own computer rather than by the remote computer). Your modem, however, will be operating with full-duplex, though the software may not be using the transmitted and the received signals at the same time.

The modem is therefore a very necessary intermediate between the computer and the telephone lines, allowing computers to be connected together so as to transfer data no matter how far apart the computers happen to be. Modern PC machines can have a modem, such as the low-cost Amstrad MC 2400, added in card form. Other modems can be connected to the computer by way of a cable from the serial port, and such a modem would need this serial connection, a mains cable connection and a telephone connection in order to work. When you use the internal modem, only the telephone cable requires to be connected into a telephone socket of the BT standard type. Beware, incidentally, of modems that are not BT approved. They may be cheap, but if BT cuts off your line because it has discovered that you are using a non-approved modem you will be charged again for re-connection. An even nastier prospect is the refusal of an insurance company to pay out after a fire if non-approved equipment has been in use.

Before you start to be too optimistic about the prospects, however, you need to know a lot more about modems. To start with, there is a huge number of modems available, mainly because there are four rates for data transmission, and very few modems, other than very costly types, in the past have offered all four. Following the introduction of the Amstrad modem, which does provide all four speeds, you can expect the prices of other BT-approved modems to be low, so that it would be rather foolish now to buy a two-speed modem on grounds of price alone. The next point is that there is a large variation in the facilities that modems, along with associated software, can offer.

The speeds that are quoted, for example, refer to the bits per second rates, and you will be surprised to find rates that are much faster than any you can use over the public telephone lines. This is because you might want to use modems for computer to computer transmissions within a building, using internal lines, or over an internal telephone exchange that can cope with higher frequencies. Even if the modem is internal you still have a serial link between the computer and the modem, and you need to know at what rates this can be used. Your software will refer to serial 'ports', meaning the connections through which serial data can be transmitted and received, and these will be distinguished by using COM1 to refer to the port connector at the back of the PC, and COM2 to refer to the internal port that is part of an internal modem. If you use an external modem, you also need to use COM1 unless you fit an internal card that provides you with the hardware for a COM2 port. Problems can arise, however, if you have two COM ports fitted and you add an internal modem (see Chapter 5).

The other facilities of a modem are less essential but nevertheless often useful. You can, for example, obtain modems that will generate dialling tones, either the musical type of tone that is used in most countries outside the UK and in modern digital exchanges operated by BT, or the slow pulses that old UK exchanges use. Both types need to be implemented at present, though all UK exchanges will eventually use the tone type of dialling. The use of an autodial modem like this permits you to have software that can store telephone numbers for you for easy error-free dialling, and eliminates the need to have a conventional telephone present in order to dial up a number. If you are working from a telephone directory of computer services, held as part of your software, then there is no risk of dialling up an ordinary voice number with the modem. In any case, when connection is made to a voice number, your modem will detect that the normal tone (put out by a modem at the other end of the line) is absent and can report this to you by, for example, printing the word VOICE on the computer screen.

Baud rates and other protocols

We have seen already that a serial transmission allows signals to be sent over a single line, such as a telephone line, and that these signals are sent bit by bit, using 7 or 8 bits per character byte along with a start bit and either one or two stop bits in each byte. The most favoured set at present is 8 data bits with one start bit and one stop bit, but you have to use whatever system is favoured by the device at the other end of the line, because there is no agreed universal standard. The baud rate that is quoted for modems or software is (in practical terms at least) the number of bits per second, so that with 8 data bits, one start and one stop bit making a total of ten bits for each character byte, a baud rate of 300 means 30 characters per second and a baud rate of 1200 means 120 characters per second.

These rates are usually referred to by modem manufacturers in a different way, using type numbers such as V21, V22, V22-bis and V23, and these designations require some explanation. The V21 system uses a speed of 300 baud for both transmission and reception, with a different tone used for each signal (0 or 1) and transmit or receive, four tones in all. This is a slow but reliable system that is well suited to noisy lines and if you are on a small branch exchange that lacks modern equipment you may be forced to use V21. If, for example, you can hear dialling noises on the line while you are making ordinary voice calls, it's likely that you will have problems if you try to make use of the faster rates. On the other hand, using V21 requires you to be on line for much longer for a given length of text, so that it is expensive in telephone time.

The V23 system is the exotic Prestel type, which gets data to you at 1200 baud, but takes your data at a leisurely 75 baud. If you join Prestel you certainly need to have the V23 available both on your modem and on your software, though in some places you may be able to make use of the more modern 1200/1200 Prestel rate. If you are confined to V23, this requires you to be careful in your selection of hardware and software, because modems and software that originate in the USA are not likely to cater for Prestel rates, and most US-designed computers (and this means all PC machines, since the original design was from IBM) cannot work with split rates. On the PC, the split rate can be achieved by the software which switches the speeds each time you change over from transmitting to receiving or vice versa. Software exists which ensures that you always have the benefit of the faster 1200 baud rate when you are transmitting, but this is not at the time of writing generally available. In any case, the faster V22 rate for Prestel is steadily becoming more easily available all over the country.

V22 is a 1200 baud rate for both transmitting and receiving, and it works on phase shift rather than on different tones. The rate of 1200 bits per second is actually achieved with a slower real baud rate, but as far as you are concerned it communicates four times as fast as a 300 baud rate so it's using the equivalent of 1200 baud. This makes it most suitable for modern telephone systems, and if you live in a district where the telephone system is good but not connected to a digital exchange, then this V22 standard is the one to plump for. There is also V22-bis, which uses 2400 baud, once at the absolute limit of what can be reasonably achieved with ordinary telephone systems and usable in the UK wherever a digital exchange is available. Faster modems can use the 9600 bits per second rate that has been common for years in many parts of the USA. The really expensive modems allow you to use V32-bis (14400 bits per second) and V42 (which uses data compression to reduce file sizes and so transfer faster). Fax modems, as the name suggests, allow you all the normal modem facilities plus the ability to transmit and receive fax, assuming that you have the software that permits this use.

Other modem features

What other modem features should you be aware of? One facility that is provided on practically every modern modem at any price range is auto-answer, meaning that the modem will accept calls automatically. In other words, while you have your computer switched on and your communications software working, an incoming call will be put through to the computer without the need for you to lift a telephone receiver. This allows your computer to act rather like an answering machine, but without your voice being heard by the caller, and it is handy if you expect data to be sent to you while you are out. The auto-answer modem will provide the correct signal to the remote transmitter to indicate that you are connected, using the same type of signal as would be sent by lifting a handset. Remember, however, that there is no point in receiving messages while you are out unless your software provides for passing such data to a disk file. In addition, if messages can be sent to you while you are out, it may also be possible for a caller to read what is on some files in your disk. This is harmless if you use floppy disks, because you can choose which disk to have in the drive, but it can be a disaster if you work with a hard disk which holds all of your programs and data. The more elaborate communications software will allow only a limited number of files to be read unless a special password is used to gain access to disk commands (commands that are often termed 'shell' commands because they are carried out by the operating system or shell, usually MS-DOS).

Another feature which appears on many, but not all, modern modems is auto-dial. As the name suggests, this allows you to carry out dialling without the use of a separate handset, and is a very useful facility even if you are only going to use a voice transmission. To be able to dial numbers automatically requires that the modem can issue either pulses (for the UK system) or tones (for other countries, UK later) and if you are buying a modem that you will be using for some time to come it's wise to get one that can use either pulse or tone, selected either manually or automatically. If the auto-dial system is described as 'Hayes' then it conforms to the standards laid down by the Hayes Corporation, the (US) leader in modem design. Using an auto-dial modem opens the door to the use of software that stores and selects telephone numbers. You can, for example, store up to 100 telephone numbers with some types of software and dial up any number simply by requesting a name. In addition, most suitable software will provide for re-dialling an engaged number (frowned on by BT) and for providing your password(s) when the other machine answers.

The third common and very useful facility is auto-detect, also called auto-scan. A modem fitted with this facility can detect what baud rate is being used by the remote machine, and set itself accordingly, so that you never have to bother about setting a baud rate for yourself. This is particularly useful if you use an auto-dial system with a lot of different numbers that use different baud rates. It does not, however, automatically set the number of data bits or stop bits for you, so that there is still something to do for yourself. As mentioned in Chapter 1, however, suitable software will allow you to keep a file of contacts that includes all of this data so that you never have to key it in more than once.

Other features are of more interest for specialized applications. Some modems allow a set of very fast baud rates for the few users who have access to specialized high-speed data transmission lines. A few types feature built-in error-detection and correction systems, but this is just as easily provided by software. Unless your needs are rather specialized, then a comparatively simple modem will be sufficient for both business and leisure communications, but simple doesn't mean crude. At one time, a simple modem would have meant V21 and V23 only, with no automatic facilities of any kind, at a price that was by no means low. At one time, a modem that offered V21, V22, V22bis and V23 with auto-receive, auto-dial and auto-detect for pulse or tone would have set you back something approaching £500. The introduction of the Amstrad modem has had the same effect on modem prices as the Amstrad PC1512 machine had on PC computer prices. You can now expect a modem to cost half as much as it did in mid-1987, and possibly less, though the Amstrad modem is no longer being advertised.

Separate modems

If you intend to buy a modem in the near future, then a low-cost internal modem for the PC, if you can find one, is still by far the best buy. You may, however, have a modem already or be attracted by a particularly good offer on a modem and its software. Such a modem is likely to be one of the external type which has to be connected to the PC by way of the serial port. If you have or are offered an old modem of the 'acoustic coupler' type, then decline politely. These devices used a microphone and loudspeaker system with your telephone handpiece placed so that the telephone microphone was over the modem loudspeaker and the telephone earpiece over the modem microphone. They were very unreliable, because any sound in the room in which they were used could upset the transmission, and the conversion to and from sound inevitably causes distortion. An even more serious handicap for use nowadays is that they work only with suitable shapes of telephone handsets, which excludes most modern patterns of telephones. Their use came about because at that time it was illegal to plug anything into the telephone system which was not supplied by the Post Office. This restriction was removed with the creation of British Telecom, and the need for acoustic couplers disappeared overnight, apart from the specialized purpose of allowing you to use a modem in a phone-box (if you know a phone-box that works) or from an hotel bedroom. You should settle only for the direct-wired type of modem which has a lead that can be plugged directly into a modern BT connector socket.

The separate modem requires three connections to be made. One of these is the connection to the BT telephone point, and this is usually made through a 'piggy-back' adapter so that you can still plug an ordinary telephone into the socket. You will also need a mains lead to supply power to the modem, with a 3A fuse in the plug. The third connector is a serial cable to connect the modem to the serial interface on the back of the PC. A cable of this type will probably be packaged with the modem and in any case is easy to get hold of because the PC serial interface wiring is as near to a standard as you are likely to find in this communications business. If you feel tempted to make up your own cable, then get a copy of the connection diagram at each end, along with a magnifying glass, a soldering iron and a set of worry beads. You'll need them.

If you have bought the cable, though, you only have to plug in at the PC end and at the modem end, put the mains plug into its socket, and the telephone plug into the BT socket, and you are ready to go if you have suitable software in the form of a communications program. The use of an external modem in this way ties up the serial port (known as COM1 or AUX1) on your computer so that you can't use if for anything else, such as a serial printer. In addition, you will need to consult your modem manual to see if any switches on the modem have to be set. Most modern modems have their settings determined entirely by the software that is used to control them, so that there are no switches (or none that you need use, at any rate) to set.

Updating the motherboard

The complete upgrading of a motherboard involves completely stripping the PC down to an empty case. The old motherboard has its earthing bolts removed, and spring clips held closed, and eased out. You can then install another motherboard in the way described in Chapter 2. Less drastic steps are upgrading memory and the use of a co-processor (not needed on 80486DX machines). Do not be tempted by offers of upgrading a processor chip, usually by way of an accelerator board which replaces the old chip, because this is often more costly than upgrading the motherboard, and leaves you with a motherboard whose performance may not be up to that of the chip. For example, if you upgrade a 386 board by fitting a 486 chip, you still have the slow bus of the old motherboard, with no chance of using a local bus. By opting for a 486 motherboard with local bus, your upgrade would be very much more effective and, if anything, less costly. At the time of writing a 486 motherboard using the 80486SX chip with local bus and cache memory was selling at around £170. Several suppliers sell motherboards and processors separately to reduce stock costs, so that if you see a 486 board offered at around £45 to £90 it is not necessarily a bargain if you need to supply the 486 chip and fit it yourself. In general, plugging in a chip such as the 80486 is not as simple as plugging in a simple 16-pin IC. The 80486 uses 168 pins arranged in three sets of rows around all four sides of the square, and locating and pressing all of these home without bending a pin is an exercise that needs practice, preferably on some dummy chips and boards.

The memory of a modern machine will be held in SIMM form, and for practical purposes we need only consider the 30-pin 1 Mbyte x 9 and 4 Mbyte x 9 SIMMs. A typical 80386SX machine will have four SIMM holders, allowing it to take up to 16 Mbyte of memory, which is as much as the SX chip can use. Motherboards using the 80386DX chip (the older versions were labelled as 80386 only) can use more memory, but few motherboards catered from more than 16 Mbyte. At the time of writing, the 1 Mbyte x 9 SIMMs are more expensive (around £52 as compared to £19 a few months previously) than they have been for many years owing to a supply shortage, but the 4 Mbyte x 9 type had not suffered such a large price increase. If your old motherboard has memory in SIMM form these SIMMs are valuable and you should either re-use them or sell them if prices are still high

You should not add memory to an existing motherboard unless you have the documentation for the motherboard, which is not always easy to achieve when you are refurbishing a machine. The older motherboards were constructed so that jumpers or switches had to be set before adding memory, and incorrect settings could cause damage to the memory chips. More modern motherboards are designed so that, provided the memory is added in the correct way, no jumper or switch settings are needed.

One thing that both older and more recent motherboards have in common is that memory is arranged in banks. This is done so as to make it impossible for the same part of memory to be addressed twice in rapid succession, and each bank is used alternately, giving the memory in a bank time to settle before it is used again. This allows slower memory chips to be used in each bank without compromising the speed of the processor. On old motherboards, this problem was dealt with by using wait states, meaning that the processor would address memory and then wait for the memory to establish its data. If you see wait states set up on a CMOS RAM, this is an indication of the use of slow memory that is not arranged in banks.

The point about using banked memory is that you cannot use odd amounts of memory. You can, for example, use 2 Mbyte, but not 1 Mbyte or 3 Mbyte, because with two SIMMs to a bank, both must be occupied to form the bank. They can both be occupied by 1 Mbyte x 9 SIMMs or both by 4 Mbyte x 9 SIMMs, but not by one of each. This means that the permitted memory sizes for a 4-SIMM machine using banking are 2 Mbyte (one bank of 1 Mbyte SIMM), 4 Mbyte (two banks of 1 Mbyte SIMM), 8 Mbyte (one bank of 4 Mbyte SIMM) and 16 Mbyte (two banks of 4 Mbyte SIMM). Remember also that you cannot mix 3-chip SIMMs and 9-chip SIMMs in a bank (and preferably not at all).

Co-processors

The Intel chip set prior to the 80386DX allowed for the use of coprocessors, chips which can take over the running of the computer from the main microprocessor. The only chip that is found to any extent as a co-processor, however, is the maths coprocessor, the old 8087 and 80287, and the more modern 80387SX. These were originally expensive chips, costing in the region of £200 or more, whose specialized action is floating-point mathematical actions. The normal microprocessor is capable of very limited arithmetical actions using integer numbers: that is, numbers which contain no fractions and have a limited range of value. Numbers which contain fractions or are outside the range of the integers that the ordinary processor chip can use have to be dealt with by the use of software, and this introduces two penalties.

One is that the speed with which these numbers can be manipulated is much lower than for integers, because each action carried out with these *floating-point* numbers (so called to distinguish them from fractions in which the decimal point is always at a fixed position) requires a large number of software program steps to carry out. During the time that each arithmetic action with a floating-point number is being executed, the main microprocessor is tied up, slowing down the overall action of the computer. The other problem is that the storage of a floating-point number in binary code form is almost always approximate, so that the results of arithmetic using floating-point numbers will be incorrect unless the results are rounded up or down and rounding requires some rather clever software instructions. A common test is an action such as

$\frac{9}{11} + \frac{2}{11}$ which will with some software not result in 1.00 but in 0.9999999999. This is a problem of binary code rather than of the microprocessor, but it adds to the amount of software that has to be executed in order to deal with floating-point arithmetic.

The maths co-processor is a chip which deals with floating-point actions using a combination of hardware and built-in software routines. The important point, however, is that it can work in parallel alongside the main processor, interrupting the main processor only when it needs to be fed with data or needs to pass results back into the main system. The use of the co-processor will therefore speed up working with floating-point numbers by an amazingly large factor. The speed improvement, however, depends on the type of software that is being used. Programs such as word processors make no use of floating-point arithmetic, so that the fitting of a co-processor to your machine will have no noticeable effect if you use the machine exclusively for word processing. If, however, you use large spreadsheets such as Lotus 1-2-3, then the effect of the added co-processor is very dramatic, obviating the annoying delays that can occur even with relatively small worksheets when every cell of the spreadsheet has to be recalculated following an entry. The same advantages will appear if you use computer-aided design (CAD) programs, or if your programs are concerned with elaborate calculations.

If you find that your use of software would justify the fitting of a co-processor, how do you go about it? The first point is that your computer must be fitted with a socket to accept the co-processor, and if this is not fitted you can forget it. Virtually all motherboards of standard design are fitted with a socket for a co-processor to match the main processor. If you make the upgrade for yourself, the co-processor chip has to be selected with some care. The type of co-processor should match the main processor, so that, for example, you need the 80387SX to match the 80386SX or the 80387DX to match the 80386DX. The 80486DX needs no co-processor because the action is incorporated into the main processor.

If you can get a co-processor that will run at the same clock rate as that of the main processor, it can be run synchronously, meaning that the actions of the two processors are synchronized. If the co-processor, as often happens, is rated for a lower speed it must be run in asynchronous mode. This means that signals have to be passed between the main processor and the co-processor to act as handshaking for the exchange of data, and this inevitably slows the process down as compared to full synchronous action. You will need to set a jumper on the motherboard when you fit a co-processor chip to specify whether the action is to be synchronous or not. Note that the 80486SX chip has its normally built-in co-processor disabled, but you can upgrade to full 80486DX action by fitting a form of co-processor which is, in fact, another 80486 with its main processor disabled.

A more recent development is co-processors from other manufacturers, notably IIT and Cyrix. These sell for much lower prices than the original Intel levels, and their immediate effect has been to cause a reduction in the prices of the Intel chips, making co-processors a much more attractive proposition to the user. The performance of these other chips is not necessarily identical to that of the Intel chips, and some care has to be taken over selection of a co-processor for any given application. In some cases, there are (minor) errors in floating-point calculations which are not present when the Intel chips are used and though this probably matters very little for applications such as CAD it can be important when floating-point calculations are used for specialist software, such as stress analysis.

Memory and program problems

Programs (or *applications*, as they are often called) are the main source of trouble for the PC user. The programs need not be troublesome in themselves, but the way that they use memory, interact with other programs and require response from the user all add up to potential trouble, particularly for the user who believes that reading manuals or books would be a waste of time. This section, then, is devoted to the memory problems that can arise simply from the use of modern programs in upgraded machines (new wine in old bottles?), and we need to start with some basic ideas and recollections about programs and memory.

The main (RAM) memory of a computer behaves like a set of switches, each of which can be set to be on or off, and each remaining as it is set only for the time during which power is applied to the computer. A set of eight of these switch-like units is called one byte of memory, and this size of memory unit is important, because this is the unit into which memory is organized, even if the computer works with larger units of two (16-bit) or four (32-bit) bytes at a time. One byte of memory is sufficient to store any character that can be represented in ASCII code, and memory sizes are always measured in units of kilobytes (1024 bytes) or megabytes (1048576 bytes). The use of the prefixes kilo and mega in this way is peculiar to computing – normally kilo means 1000 and mega means 1000000 – and the reason for the difference is that the memory 'switches' can be set one of two ways, so that numbers have to be stored in twos. 1024 is 2 to the power 10, so that this is preferred to the use of 1000 for computing purposes.

The microprocessor of the computer can gain access to any byte of memory by a very simple system of numbering each byte and using the number, called the address number, as a reference. In order to gain access to memory, the microprocessor must place signals corresponding to the address number on to a set of lines called the address bus, and there must be memory physically present which responds to the address. It's all very simple and logical so far, but this is where the problems start. The original 8088/8086 processors that were used on the original IBM PC machines (and the many clones that were constructed later) were capable of addressing 1 Mbyte (1024 Kbyte) of memory, an amount which at the time was thought to be ridiculously excessive. This does not mean, however, that all of this 1 Mbyte of memory can be RAM that can be used by programs. The machine needs some memory to be present in permanent (ROM) form, and some address numbers need to be reserved for this section of the memory, particularly since the microprocessor chip is made so that when it is first switched on or reset, it will try to read memory at a specific address, number 1048560. This is built into the chip and cannot be altered, so that any computer using the Intel chips must provide for the BIOS program to start at this address – the remainder of the routines can be elsewhere as long as the start is at this fixed address.

A further complication is added by the use of the video graphics card. The video signal depends on using memory, and the PC type of machine reserves memory addresses for this purpose, the video memory, with the RAM chips for this placed on the video graphics card itself. The amount that is needed depends on the type of video card that is being used, with SVGA requiring up to 1 Mbyte, EGA and VGA requiring 256 Kbyte, MDA least (see Chapter 1 for details of these cards). Finally, the MS-DOS operating system, certainly in versions prior to 4.0, can use only 640 Kbyte of RAM for programs. What this all amounts to is illustrated in Figure 7.5 in the form of a memory usage diagram. The memory from 0 to 640 Kbyte is RAM which is described as system memory or base memory. This is used for the MS-DOS operating system itself, for other resident programs (see later), and for the programs (applications programs) that you want to use. The region between 640 Kbyte and

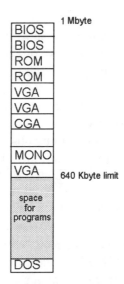

Figure 7.5 *Typical memory use. The amount taken for DOS depends on how the computer is set up, and MEMMAKER will optimize this. Memory above the 1 Mbyte mark can be used only by programs that have been designed to use this memory, and such programs are incompatible with old PC models.*

768 Kbyte is reserved for video boards. The EGA and VGA types of boards will carry memory chips corresponding to the whole of the permitted range, MDA, Hercules, and CGA cards will need less and will carry fewer memory chips. The region between 960 Kbyte and 1024 Kbyte is reserved for BIOS ROM. This region consists of 64 Kbyte of address numbers, and most BIOS chips use no more than 16 Kbyte, but the remainder is not entirely free for use. What is to some extent free for use is the range of memory addresses between 768 Kbyte and 960 Kbyte, some 192 Kbyte of addresses. Even some of this is likely to be used, and if you have a hard disk and/or a VGA or EGA card, some of this set of addresses will be used by the ROMs that are required for these systems.

What it all amounts to in practice is that the 640 Kbyte of RAM memory starts to be strained when you use large programs of around 360 Kbyte or so, because when you start the computer, a chunk of the MS-DOS system remains permanently in the memory, taking up some of the lowest part of the memory and a small portion of the uppermost piece of the 640 Kbyte of RAM. MS-DOS versions earlier than 4.0 take up about 70 Kbyte of memory (version 4.0 requires considerably more, but later versions 5.0, 6.0 and 6.2 need less), and there are always other resident programs such as the country files which also take up memory from the absolute limit of 640 Kbyte. Most programs need some RAM to use with data, and the effect of all these calls on memory is to limit the size of a program to the 360 Kbyte or so mentioned above. MS-DOS versions 5.0 and 6.0 allow more of the base memory to be used on modern machines with extended memory by placing some resident programs and parts of the MS-DOS system itself into memory addresses beyond the 1 Mbyte mark.

More memory

The PC/AT type of machine nowadays uses the 80386 or 80486 type of processor which can address very much more memory than the 1 Mbyte limit of the 8086/8088 chips used in the original PC and XT machines. The 80386SX machine can address up to 16 Mbyte, and the 80386DX and 80486 up to 4096 Mbyte. Even the most sprawling of modern programs will fit in these limits. Such machines often come with at least 2 Mbyte of RAM on the main board (the motherboard) and provision for adding more memory, usually 4 Mbyte or 16 Mbyte, on the motherboard. This type of memory above the 640 Kbyte limit is called extended memory. Unlike the expanded memory which could be added to the older machines, extended memory is addressed directly with each byte of memory having its own unique address rather than the expanded memory system of using a 64 Kbyte set of addresses to use 64 Kbyte pieces of memory one at a time. Programs designed to run under MS-DOS, however, do not make use of these memory address numbers above the 640 Kbyte limit (with a small exception of 64 Kbyte above the 1 Mbyte mark). To do so would make the programs incompatible with the older XT machines which are physically incapable of generating these address numbers.

In an AT machine with 1 Mbyte of RAM, this first 1 Mbyte is usually split with a lower section using addresses up to 640 Kbyte, and the remaining 384 Kbyte located at addresses just above the 1 Mbyte point. Any further extended memory is located in the following addresses, joining seamlessly to the 384 Kbyte from the first megabyte. This makes the first megabyte of RAM of an AT machine particularly important. The addresses between the 640 Kbyte mark and the 1 Mbyte mark are used in exactly the same way as they are in the old XT machines in order to maintain compatibility. Unfortunately, the parts of this memory above the 640 Kbyte limit cannot be addressed when the machine is using only MS-DOS. Extended memory can be used by programs if the rules laid down by Microsoft are followed, using a memory manager system called XMS. Very few programs, however, currently make use of extended memory in this way, but two important exceptions are Lotus 1-2-3 V.3 and Windows 3.1.

Using Windows 3.1 (see Chapter 9) allows extended memory to be fully used by any of the programs that are being run under Windows. The new operating systems, NT and Windows 95, that Microsoft have developed will be able to replace both MS-DOS and Windows in this respect, but will run only on 80386, 80486, Pentium and later (yet to be announced) machines. In short, unless you use Windows, the extended memory of your computer cannot at present be used by programs. Extended memory has to be controlled by a memory-manager program.

The only applications open to you if you do not use Windows or another program that makes direct use of extended memory are RAM-drive, cache memory and printer buffering. RAMdrive uses a manager program that treats some of the memory as if it were a disk drive of very high speed (but which loses its information when you switch off). Printer buffering requires a program which uses the extra memory as an intermediate between the computer and the printer, so that the computer is not tied up waiting for the printer to finish. Cache memory, which also needs a manager such as the MS-DOS Smartdrive, uses memory as an intermediate between the computer and the hard drive, so that the hard drive is used to a lesser extent, with data being supplied from the (faster) memory until another access to the hard drive is required.

Programs and memory

A program is a set of instructions to the microprocessor, consisting of a set of number codes that the processor must read in sequence. These number codes cannot easily be distinguished from the ASCII codes of text, but if you try to print a program file on your screen (using the TYPE command of MS-DOS) you will see mainly gibberish, though the odd sensible word or phrase will appear if the program contains instructions. There is a another vital difference between a program file and a file of text. If a file of text contains an error it does not make the file useless – you can use a word processor or editor to correct the mistake, or even simply put up with it. An error in a program file is much more serious. At best the program will be 'brain-damaged', it will certainly not work as intended in some respect. At worst, the whole program will run amok, corrupting the memory and endangering resident programs and even, in some cases, disk files.

If you suspect that a program is misbehaving and there is a chance that its file is corrupted you should stop using the program at once, restart the computer, and replace the existing version with a back-up version. The normal behaviour of programs is far from simple, and this is no place for a full explanation, but some aspects need to be understood in order to see why things work out the way they do. A program that uses the keyboard and the screen, for example, ought to do so by making use of the MS-DOS system. The operating system exists to provide standard routines that programs can harness to their own requirements. If every program made 100% use of the MS-DOS system for all its use of keyboard, mouse, video display, memory and disk actions then any computer which uses MS-DOS could run any program.

This is, unfortunately, not true, and though most programs use the MS-DOS disk routines, and a very large number use the keyboard and memory-management routines, few use the video display routines of MS-DOS. This is because a large number of programs which use the screen need to make fast changes to the screen display or to draw graphics on the screen. The use of the MS-DOS routines is a distinct handicap for such actions, and so virtually all programs which make intensive use of the screen (word processors, spreadsheets, CAD, DTP and other graphics programs) adopt one of two possible ways around the use of MS-DOS. One of these is to use the routines that exist in the ROM BIOS – in many cases, calling a routine in MS-DOS simply results in one of these ROM routines being called anyhow, so that calling the ROM routine directly saves time. This implies that the program can be used only on a machine whose ROM is standard, as compared to the IBM ROM, meaning that the starting addresses for the routines in the ROM will be at the same addresses as in the IBM version (many machines will, of course, use a ROM licensed from IBM).

The other option is a bolder one, to make direct use of the 'video addresses' in the video card, the addresses in the region 640 Kbyte – 768 Kbyte. This is less easy, because it means that the program must contain all the routines for manipulating the video memory, rather than using the ready-made routines of MS-DOS or ROM, and this requirement makes a program considerably longer and more complicated. The speed, however, can be very much higher, because even a direct call to a routine in ROM is slower than using a routine in RAM, since ROM is a slower-responding type of memory. Most modern computers can shadow ROM, copying ROM into RAM (using the same address range switched to RAM) in order to make faster use of the ROM routines. If this latter method is used, then the video addresses and the way that the screen display is organized must also be identical to the methods used on the IBM machines. To be compatible, then, a computer must use the same addresses for routines in the ROM as IBM and must also use the same addresses in video memory, with the screen organized in the same way.

This is why it is possible to find computers which run MS-DOS but which are not IBM-compatible and which will not run the huge range of programs that an IBM or true compatible machine will run. Non-compatible MS-DOS machines have almost died out, but a few can still be found offered for sale at 'bargain' prices – one useful indicator is that they often use a very old version of MS-DOS such as version 2.1.

TSR programs

TSR programs are loaded into the memory and remain resident there. Unfortunately, little consideration is given to ways of removing such programs, and this leads to two problems. One is that if a TSR program is loaded each time another program is started, frequent starting of this program will result in several copies of the TSR being loaded, taking up too much memory. The other point is that some TSR programs may interfere with other actions, so that they become an unwanted nuisance once the program that requires them has ended. This is mainly a problem for programs running under MS-DOS because the better memory management of Windows can cope with TSR uses. At the time of writing, a very few applications provide for stripping out TSR programs which they load, and if you need to be able to strip them out, a pair of Public Domain programs called MARK and RELEASE provide a simple method. If the MARK program, itself a TSR, is loaded in prior to any other TSR, running RELEASE will remove both MARK and the TSR.

For modern machines running Windows 95, the problems of de-installation have been tackled, but only for the programs that have been written specifically for Windows 95. These programs are installed using the Control Panel of Windows, and they can be de-installed in the same way, removing all sections of code that belong to the program.

Section III

The proof of the pudding

Chapter **8**

MS-DOS – the
operating system

NOTE: Throughout the following chapters which deal with essential software, the key that is marked with the large arrow, Figure 8.1, or (unusually) labelled as Enter or Return, will be referred to as the *Enter* key.

Figure 8.1 *The Enter or Return key, which is normally of a distinctive shape and labelled with this arrow.*

MS-DOS is an operating system, a program that either exists in the computer in the form of a chip (a ROM or Read Only Memory system, used now mainly for portable machines) or which can be loaded in from a disk, using a very small loader (or boot) program held in the ROM. For a machine that you have constructed for yourself or upgraded, the MS-DOS files will be on a disk or a set of disks. The aim of the operating system is to provide, at the very least, the essential routines that every computer needs to control:

- The screen so that what you type can be seen on the monitor

- The keyboard so that you can type commands

- The disk drive(s), for managing files and loading and running programs

- The printer port and serial ports

Without program routines that attend to these tasks, the computer is usable only if each program that is to be run attends to all of these actions for itself. This would be a wasteful duplication, so that the operating system is just as important as all the hardware of the computer – a PC machine with no operating system simply cannot be used. Having the operating system on a disk allows the operating system to be improved over the years, something that is much more difficult for a system that is frozen in the form of a ROM chip.

When microcomputers started to be used seriously for business purposes, many manufacturers devised their own operating systems (OS). The use of different operating systems by different manufacturers, however, makes it virtually impossible to set standards that will allow one computer to run a program that works on another computer from a different manufacturer. In these early days (1970-1980) it was particularly important to be able to control the action of computers that used disk drives, so that a Disk Operating System, or DOS, was essential. One solution to the problem of interchangeability came about in the very early days of the development of microcomputers, in the form of CP/M. This was a system to control, program and monitor (hence the initials) the operation of these early microcomputers, and it was devised by Gary Kildall in or around 1973, before microcomputers were a commercial proposition and when the idea of a standard operating system for microcomputers was quite new.

CP/M came provided with a set of utility programs that allowed the user to carry out tasks like file copying, file checking, program modification, disk verification and other housekeeping actions that are now so familiar. When the original IBM PC was developed, the small firm of Microsoft offered an operating system that allowed more control and more logical commands than CP/M. This started the remarkable growth of MS-DOS and of Microsoft itself. Since these early days, MS-DOS has become the standard operating system for the PC form of machine. This standardization has been one of the attractions of the PC, because it has ensured that upgrading the computer did not require upgrading all of the software. This situation is changing, because modern machines are handicapped by the requirement to be able to run programs that could be run on the original IBM PC. The next few years will see considerable steps in the provision of more modern PC operating systems that still retain compatibility with older software.

MS-DOS has grown from a small operating system, used mainly by programmers, to a set of commands that are used intensively by millions of computer users all over the world. Though Microsoft itself has developed more convenient ways of controlling a computer in the shape of Windows, the MS-DOS system currently remains in charge, because Windows 3.1 simply makes use of MS-DOS. This means that if you can use MS-DOS directly you can exercise tighter control and make your commands run faster than is possible when there is another program in charge of MS-DOS. At the time of writing, fewer users work with MS-DOS directly, and what follows is intended for owners who have upgraded to 80386 with 2 Mbyte of memory and running DOS programs, which are still plentiful and cheap. If you use Windows 95, you have very little contact with MS-DOS (which is still present), and you need not spend much time on details of MS-DOS.

We can group the actions of MS-DOS into sets:

1   File actions, which allow you to control the files of programs and data that are held on the hard disk

2   Disk actions, which allow you to format and use floppy disks, and which organize the hard disk for easier use

3   Utilities, which are designed to add convenient methods of carrying out actions that at one time would have required you to be able to program the computer for yourself

4   More advanced methods of controlling the use of memory and the hard disk

It is important to realize that MS-DOS totally controls the PC machine. Many machines now use the Microsoft Windows system along with programs written specially for Windows. Though convenient, Windows does not replace MS-DOS, and it has to use MS-DOS to carry out its actions. This is rather like asking someone else to use the remote-control for your TV – it might be less effort, but it takes longer, and you are giving up your ability to control things directly.

Files

Files on a floppy disk are identified by a filename. This can consist of three parts: the drive or path, the main filename, and the extension. A suitable main filename will consist of up to eight characters, the first of which must be a letter. The other characters can be letters, or you can use the digits 0 to 9, or the symbols:

$ # & @ ! % ( ) – _ { } ' ~ ^ '

The following characters must not be used:

* + = [ ] ; : , . / ?

Nor can you use the space, the tab, or a Ctrl character as any part of a filename.

Unless it's particularly important for you to use these permitted symbols, you should avoid them, particularly symbols like the single inverted quotes, which are easily confused or overlooked. A good rule is to use recognizable words with a digit or pair of digits used at the end to express versions, like TEXT001, TEXT002, ..., TEXT015 and so on. Allowing up to eight characters means that you may have to abbreviate names that you would want to use, like ACCREC1 or BOTLEDG1, but it should be possible to provide a name that conveys to you what the file is all about. Whether you use lower-case or upper-case characters, MS-DOS will convert all filenames to upper-case (capital letters). Windows 95 allows you to use longer filenames, even including spaces.

The eight-character (or less) main name for a file is essential, but there are two other parts to a filename that are often optional. One of these can be a drive letter, usually A:, B: or C:, or a path (see later); the other is the extension, a set of up to three letters. The drive letters A: or B: will be used only if the program is on a floppy disk. For example, if you have a program called SPLAT on a floppy disk that is in the A: drive, you can run the program in either of two ways:

1   With the prompt appearing as C:\>, type A: and press the Enter key. When the prompt shows as A:\>, type the program name of SPLAT and press Enter. When the program finishes the prompt will be A:\>.

2   With the prompt appearing as C:\>, type the command A:SPLAT and press the Enter key. When the program finishes the prompt will be C:\>.

These methods allow you to run a program no matter where it is located, and the second method allows you to return to the hard drive after running a program from a floppy disk. The first method starts by changing to the floppy drive, so that the floppy drive will be the current drive after the program has finished. This scheme can be extended to using directory names as well as a drive letter (see later).

The extension is a set of up to three letters that are added to the end of the filename, separated by a dot. Any letters following three will be ignored when the filename is typed. Like the disk drive letter, the extension is optional, and if it is not used, then the dot is not needed. This is why the use of the colon or the dot is prohibited within an 8-character main filename.

The purpose of the extension is to convey some extra information about the type of file, though you can make whatever use of it suits your own purposes, within limits. Just as there are forbidden filenames, there are also forbidden extensions, and in particular you should not make use of the extensions .EXE, .BAK, .HEX, .COM or .OBJ for your files, because these are extensions that are used with special meanings. The list following shows a few 'standard' extension letters and their uses, and you should, if you use extensions at all, either keep to some of these or use some entirely different codes of your own.

BAK	A back-up file
BAS	A program that cannot be run unless BASIC has been loaded first
BAT	A batch file of commands to the DOS
COM	A program that can be loaded and run by typing its (main) name
DOC	A text file of documentation for a program, or a file produced by Microsoft Word
EXE	As for COM, but a longer program
MSG	A text file of instructions
SYS	A file that is used by the operating system
TMP	A file that is created temporarily and wiped later
$$$	Also a temporary file

A filename whose extension is COM or EXE is the name of a program file, and the program will run when you type its name – you do not need to type the extension letters. For example, if you are using a directory that contains a program file called WRITIT.EXE then you can run the program by typing:

```
WRITIT
```

and pressing the Enter key.

The same applies to files with the COM extension, and the more specialized type with the BAT extension (see later). The BAT type of file is not so much a program file as a file that runs programs.

Some programs require you to add other information, called parameters, on the command line (meaning that you type them following the name of the program). These parameters are always separated from the program name by at least one space, and they sometimes require you to use characters like the slashmark (/), which must not be confused with the backslash (\).

Directories

Another way of ensuring that files do not conflict is using directories. A directory is a group of files, usually on the hard disk, but you can create and use directories on a floppy disk as well. The principle is to keep sets of files together with the advantages that it is much easier to locate files on a hard drive, you can have two files with the same name co-existing provided they are in different directories, and you can obtain a list of files for one directory by itself, as if it were a separate disk. The drive name, such as C:\, with no directory name added, is called the *root directory*.

A hard drive may house several thousand files. This would make it very difficult to find the file you wanted unless they were grouped, and you would also be subject to the usual rules that saving a file would always replace an existing file of the same name. In addition, all disks have a limit to the number of filenames that can normally be used in the root directory. This is 128 for a hard drive root directory, but a subdirectory full of files counts as just one name in the root directory, so that if you have 1000 files in a directory (not recommended), this counts as just one name in the main (or root) directory. By using directories you can obtain a list of all the files in one directory, usually a manageable amount, rather than all the files on a disk, which could be a very large list.

There is no limit to the number of files you can place in a directory, other than the capacity of the disk itself. Each directory is named just as a file is named, with a set of letters, up to eight. You can then locate a file by naming its directory in the way that you would name the drive if you were using the A: or B: drive. Suppose, for example, that the SPLAT program is located in the C:\ODDPROG directory, and you are currently using the C:\ root directory. You can then run the program by typing, as a name:

```
C:\ODDPROG\SPLAT
```

The C:\ODDPROG portion of this full filename is called the directory path. This full directory path is required if you are not using, as your current directory, the directory that contains the program you want to run.

Hard-drive formatting

A new hard drive must be formatted, unless it has been supplied completely formatted, so that the essential MS-DOS tracks are placed on it, before it can be put into service. The type of formatting that is needed is called *high-level formatting*, and it must not be confused with the low-level format that was used for the older types of drives. The high-level format of a hard drive follows much the same pattern as the format of a floppy disk, establishing tables that will be filled with the data on files (start and end positions of each block and total length) when the disk is used for files. The system that is used establishes a file allocation table (FAT) and a directory, of which the operating system uses mainly the FAT. The data stored in the directory can be read by using an MS-DOS command, but the FAT can be read only by specialized diagnostic software.

Formatting a new hard drive requires a floppy MS-DOS system disk, the first disk of a set that holds all the MS-DOS files. If you have never used any version of MS-DOS before, you need to buy a master copy of MS-DOS in a form that can be used to format a new drive and install MS-DOS. This is an important point, because some versions of MS-DOS were sold either as an upgrade, meaning that they could be installed only on a machine that was already running MS-DOS, or as a new system for machines with no existing DOS. The most recent version of MS-DOS, version 6.22, can be used to format the new hard drive and will then install all the other MS-DOS files.

If you are upgrading a machine you will already have at least one MS-DOS master disk in stock, but if it is of the wrong type (a 5¼" disk, for example, with the new machine needing 3½" disks) you will need to borrow a system disk on the correct disk type, or buy a new set of MS-DOS disks that can be used to format a hard drive. The following description shows how to format the hard drive with system tracks from such a floppy disk.

Place the MS-DOS system disk in drive A, and start the machine. If you have not already set up the CMOS RAM, do so, and make sure that booting from a floppy is permitted. If you have set up the CMOS RAM previously, the machine should boot from the floppy. Wait until all disk activity has ceased. You will then see the characters that are referred to as the prompt, in this case A:\>, meaning that the A: drive is in use and the machine is waiting for a valid command.

With the system disk in floppy drive A:, check that it has the format command on it by typing `DIR FORMAT.*`. If you see a line such as:

```
FORMAT.COM
```

appear, you know that the FORMAT program is on the disk. If it is not, find which disk this program is on.

When you find FORMAT.COM, with that disk in the A: drive, type the command:

```
FORMAT C:/s
```

The effect will be to format the hard drive and place the MS-DOS system tracks on it, so that booting will in future be from the hard drive. Remove the floppy disk and press Ctrl-Alt-Del to check that the machine will now boot from the hard drive. Other MS-DOS files can then be copied to the disk as required.

If you have bought a set of MS-DOS 6.22 upgrade disks, and your hard drive is unformatted, place Disk 1 of the set into the A: drive, and type FORMAT C:/s as noted above. Once this has been done, remove the floppy from the drive and check that pressing Ctrl-Alt-Del will cause a boot from the hard drive when the floppy disk is removed.

Either way, you can now restart, select CMOS RAM Setup and alter the boot order so that the preferred order is C:,A:, and with floppy seek disabled.

MS-DOS 6.0, 6.2 installation

The procedures mentioned above will format the hard drive and place the essential tracks on it that allow it to be used as a boot disk. Only a few MS-DOS commands are installed, however, and though these include the essential commands you might need to use others. Unless you know the MS-DOS system well it is difficult to be certain about which files are useful and which will slumber on the hard drive until the day it is scrapped, so that it is better to install the system fully and delete files as and when you feel they are of no use to you.

Installation of MS-DOS 6.22 on a formatted drive is done as follows.

Lay out, in order, the floppy disks that are used to distribute MS-DOS 6.22. The standard set consists of three 3½" high-density disks, though 5¼" high-density disks can be supplied if requested.

It is important to realize that you cannot install MS-DOS 6.22 simply by copying files to your hard disk, because the files are stored in compressed form, and the SETUP utility should normally be used to expand and place these files, though they can be manually expanded and copied if required. The compressed files are marked by using an extension that ends with an underscore, such as MSDOS.SY_ or UNIN-STAL.EX_.

Note that when a pause is used during installation to allow you to read a notice, the Enter key is used to continue installation, the F1 key to obtain help, and the F3 key to exit.

Now insert the first MS-DOS distribution disk into drive A:. Switch to this drive by typing A: (press Enter), and then type SETUP (Enter). The SETUP program will check your system, and remind you to have one or two UNINSTALL disk(s) ready. Remember that you must place these into the A: drive when instructed by a screen message. These disks are a backup system that is used when an existing version of MS-DOS is being replaced, so that they will contain nothing useful when you are starting with a new hard drive that is only formatted.

Wait until any messages about diagnostic reports are completed. You are then given the opportunity to continue with SETUP (press the Enter key), to learn more (press F1) or to abandon SETUP (press F3). You will again be reminded of the need to format and label a 3½" 1.44 Mbyte or 5¼" 1.2 Mbyte disk as an UNINSTALL disk. You will need to prepare two disks if your A: drive (unusually) uses a 360 Kbyte or a 720 Kbyte drive. Press the Enter key to continue with SETUP.

You will see on screen a report on the existing operating system software and video hardware configuration that you can confirm or change as required. This should show the entry for DOS PATH (the directory where MS-DOS utilities are stored) as MS-DOS or DOS. If this is not so, type in this name. If you are upgrading an old version of MS-DOS, the directory is called C:\MSDOS, but a new directory created by SETUP will usually be named C:\DOS. Note this name, because you will have to use it later.

You will see on screen the options of Backup, Undelete and Anti-Virus and you are asked which program options you want to install. By placing the cursor on any of these options you can install any combination. The MS-DOS 6.22 SETUP provided for installing Windows versions of these utilities, but only if Windows has already been installed on the computer. You will also be offered the installation of the DRVSPACE utility which allows you to make more efficient use of your hard drive. You are reminded that SETUP should not be interrupted from this point onwards, and are given a final chance to leave (press F3) or to proceed (press the Y key).

The process is automatic from this point. You will be asked to change disks at times. The first of these times occurs when you are asked to place the UNINSTALL disk (or UNINSTALL #1 if there are two) into the A: drive and press Enter to notify this. The progress of installation is marked by a display that shows (on a colour monitor) as a yellow bar on a blue background. This bar lengthens to indicate the extent of installation. Several messages appear during installation to remind you of facilities that will be available when you have completed installing MS-DOS 6.22. When UNINSTALL is complete, you are asked to return to the first SETUP disk and press Enter.

The SETUP program will request you to insert the second disk of the 1.44 Mbyte set after about 33% of installation has been completed. The other disks will be requested in turn. While the disks are being read, you will see messages on the screen about the benefits of using the MS-DOS 6.22 utility programs. When the last disk has been read, you are asked to remove all floppy disks and press the Enter key. You are reminded that MS-DOS 6.22 is installed, and that old file versions have been saved on the UNINSTALL disk(s). Pressing the Enter key again will restart the machine running MS-DOS 6.22 (it will have been running the old DOS up to this point). You should see the screen prompt now appearing as C:\>.

Making system floppy disks

It is also important now to make several SYSTEM floppy disks from which you can boot the computer if there is at any time a failure of the hard disk. You must make copies also of the files that are called CONFIG.SYS and AUTOEXEC.BAT and, if you use the MS-DOS DriveSpace utility, the file called DRVSPACE.BIN. You can use new unformatted blank floppy disks or blank formatted disks as you please – the process takes longer if you use unformatted disks. Make at least two such disks now.

Place a new unformatted floppy into the A: drive. Make sure that the prompt on the screen reads C:\DOS>. If it does not, then type:

```
CD C:\DOS          (press the Enter key)
```

This assumes that the MS-DOS files are in a directory of the hard drive called C:\DOS. If they are in a directory called C:\MSDOS, use this name. Now type the command:

```
FORMAT A: /S       (press the Enter key)
```

Note the space between T and A, and also between the colon and the forward slashmark. You will be asked to insert the floppy and press Enter – since the floppy is already in place, simply press the Enter key. Wait until the process is complete before you remove the disk – the floppy disk drive light should be out and there should be a message on the screen asking you if you want to format another disk.

Insert another new unformatted disk, press the Y key to answer the question about formatting another disk, and then again to acknowledge that you have a disk in the drive. Wait until this disk also is formatted. When the message about formatting another appears, answer with the N key. You should see the C:\DOS> prompt message again.

Keep this disk in place and type the following set of instructions, pressing the Enter key after typing each instruction:

```
COPY C:\AUTOEXEC.BAT A:
COPY C:\CONFIG.SYS A:
```

This assumes that COMMAND.COM is put on to the disks by the FORMAT command, as is normal. You can check this by using the command:

```
DIR A:
```

and looking for the name COMMAND.COM in the list. If it does not appear, then use:

```
COPY C:\DOS\COMMAND.COM A:
```

(assuming that COMMAND.COM is in the C:\DOS or C:\MSDOS directory of the hard drive, which is normal. If files are not in their usual places, you will have to find where these files are located, and type the correct directory name into the COPY line. When the C:\DOS> prompt returns after copying the files, and the floppy disk drive light goes out, remove the disk and insert the other one, then repeat the step above. When this disk is ready, remove it from the drive. Label both disks as MS-DOS 6.22 SYSTEM DISK (for example), set the write-protect tabs, and store the disks in a cold dry place well away from magnets. If you use the DriveSpace utility of MS-DOS 6.22 you should also ensure that the file DRVSPACE.BIN is also copied to the floppy.

These disks are your safeguard in the event of anything unfortunate happening to the hard disk. You will normally start (boot) the computer with no floppy disks in the drives, allowing the hard drive to provide all the files that are needed to start and control the machine. If anything should go wrong with the hard drive you will get a message that calls for you to place a SYSTEM floppy in the A: drive – and that's when you will be glad that you made these disks.

A hard drive is a reliable device, but it is a mechanical device and like all mechanical devices it will eventually fail. You do not necessarily get any warning of this, which is why it is essential to keep SYSTEM disks and to make backups of all your valuable data. System disks are also your front line of assistance in the event of contracting a virus.

Summary – running programs

1　You can change from one drive to another by typing the drive letter followed by a colon and pressing the Enter key.

2　You can run a program that is stored in the current drive or directory by typing the main program name and pressing the Enter key.

3　You can run a program that is stored in any other drive or directory by typing the drive letter (along with directory path, then the program name, and pressing the Enter key.

4   You never need to type the extension letters of the program, and you must not type the full-stop that separates the extension from the main name.

Error messages

MS-DOS can deliver a huge range of error messages when a program action is, for some reason, impossible. There is no space in this book to list the error messages, some of which are delivered in circumstances so unusual that the phrase 'blue moon' seems appropriate. Many of these error messages are obvious, such as having no floppy disk in the drive that you have specified, or no file of the name you have typed. Others are obscure unless you know how the system works. In addition to the MS-DOS messages, you can get error messages from programs, with each program delivering its own set of error messages which should be reasonably obvious and which will be listed in the manual for that program.

A full list of MS-DOS error messages is contained in the Newnes *MS-DOS Pocket Book*, and a few of the most common but less obvious are listed below, together with one error message that can be delivered from other software and which can cause some confusion. No attempt has been made to explain in detail, because there is no space to do so, and the explanation is enough to allow you to find out more from other books such as the *MS-DOS Pocket Book*. Remember that it is quite unusual to get some of these error messages.

Allocation error, size adjusted
There is a discrepancy between the amount of data in a file and the file size as recorded in the directory. Use CHKDSK /F to find any stray pieces of the file.

ANSI.SYS must be installed to perform requested function
You have tried to use a screen action that needed the DEVICE=ANSI.SYS line in CONFIG.SYS. This is very unusual nowadays.

Attempting to recover allocation unit
Message from FORMAT when a bad allocation unit has been found.

Bad command or file name
Check your typing using F3 (or DOSKEY commands); there is something wrong (spelling, syntax) with the command.

Bad or missing Command Interpreter
The COMMAND.COM file cannot be found. On a floppy-only machine, insert the system disk into drive A. On a hard disk machine, check directories; absence of COMMAND.COM may mean severe disk corruption or lack of a PATH statement in AUTOEXEC.BAT.

Bad or missing filename
A line in the CONFIG.SYS file is incorrect. A device name may be misspelled or its SYS file missing.

Bad or missing Keyboard definition file
The KEYB command has specified a KEYBOARD.SYS file that cannot be found; check the directories and/or path in use.

Cannot find system files
The main IO.SYS and MSDOS.SYS files cannot be found. Insert a system floppy and reboot, then check the hard drive for serious disk corruption. Another cause is an exhausted CMOS RAM battery. These files are not seen in a DIR listing, and you can check their presence by using DOSSHELL or Windows, with the Files options set to allow viewing of hidden/system files.

Cannot load COMMAND.COM system halted
COMMAND.COM cannot be found. Use a system disk and reboot. If COMMAND.COM has disappeared from the hard disk check for serious disk corruption.

Cannot read file allocation table
The disk is seriously corrupted; you may be able to use utilities to rebuild the table, but this does not solve the problem of why the corruption should have happened. Check the disk surface using a suitable utility.

Cannot start command-name, exiting
Possible causes are too low a value for the FILES= line in CONFIG.SYS, path to COMMAND.COM incorrect (put into MSDOS directory and ensure correct path) or insufficient memory to load COMMAND.COM (very unusual).

Current drive is no longer valid
You are using either a network or an empty floppy drive (no disk, or door not closed).

Data error reading drive X:
Defective disk, often unformatted or created by an incompatible machine. Applies also to disks created by some backup programs, which are formatted in a special way.

Disk error reading/writing drive X:
Defective, unformatted or incorrectly formatted disk.

Disk error reading/writing FAT X:
Defective file allocation table; MS-DOS is using spare table. Copy files to another disk as soon as possible. Run CHKDSK/F on the faulty disk.

Duplicate file name or file name not found

You cannot rename a file to a name that already exists, and you cannot load or copy a file that does not exist.

Error loading operating system

Boot from hard disk is faulty. If this persists, boot from a floppy (you do have several spares, don't you?) and use the SYS command to put a new copy of MS-DOS on the fixed disk.

Error reading directory

Bad sectors on disk affecting directory or file allocation table. If you have backup copies, discard a floppy disk with this error or reformat a hard disk with this error. If you have no backup, use utilities to recover as much as possible.

Error reading MS-DOS system file

No system file on floppy disk, or corruption of system files. Boot from another disk, and check suspect disk.

File allocation table bad

Defective disk; use CHKDSK/F.

File allocation table bad drive X:

Disk not formatted or incorrectly formatted (or not using MS-DOS). Try CHKDSK/F, or reformat.

File MSDOS.SYS not found on specified drive

A disk which is being used as a system disk lacks this essential file.

General failure reading/writing drive X:

Incorrectly formatted disk, or failure of drive.

HMA not available: loading DOS low

CONFIG.SYS contains DOS=high, but no high memory is available. Check that device=himem.sys is used before dos=high, and that memory is not allocated to shadow RAM.

Incorrect DOS version

You need to use the version of DOS for which the program was created (see SETVER for MS-DOS 6.22).

Infinite retry on parallel printer timeout

Printer off-line, out of paper, not ready or working very slowly. This may hang up the computer unless the software allows some way of releasing it.

Insert disk with COMMAND.COM in drive X:

Put system disk in drive; this message should never be seen when a hard disk is being used.

Internal error

Usually an error within a utility, no simple cure. Try another copy or version.

Non-system disk or disk error
Replace and strike any key when ready
Put a system disk into the drive and press any key. This message should not appear when a hard drive is in use, and if it does appear, the hard drive is corrupted or the CMOS RAM backup battery is exhausted.

Out of environment space
No more SET commands can be used because of memory restrictions. Use /E with COMMAND to create more space, or edit out excessive SET commands.

Probable non-DOS disk
Continue (Y/N)?
Disk may have been created by another version of MS-DOS or a different type of DOS. Disk probably unusable.

Specified command-name *search directory bad*
An incorrect SHELL line in CONFIG.SYS.

Syntax error
A command is wrongly spelled or incorrectly used.

Too many open files
Edit CONFIG.SYS to increase the number in the FILES= line, and try again.

When other software delivers a message to the effect that COMMAND.COM cannot be loaded, this is an indication that there is corruption of this file, or that the AUTOEXEC.BAT file contains no suitable PATH or SHELL line. Check that the PATH line contains both C:\ and C:\MSDOS (or C:\DOS), and that there is a line starting with SHELL (which should have been put in automatically by installing MS-DOS 6.22) which specifies a path to COMMAND.COM. COMMAND.COM contains the MS-DOS command codes, and when you run a program, most of COMMAND.COM is removed until the program ends, so as to give the maximum amount of memory to the program. The missing bits of COMMAND.COM are re-loaded when the program ends, and the error message is delivered if the COMMAND.COM file cannot be found or is corrupted.

Diagnostic software

A very important weapon in fighting PC misbehaviour is the use of diagnostic software. There is a large range of software that can be used for all sorts of purposes, from checking floppy disk drive rotational speeds to the use of memory, but unless you have a pressing need for specialized software, a general-purpose diagnostic is more useful. In particular, the MSD program that is packaged with MS-DOS 6.22 is useful because it can provide an output in disk-file form that can be sent to a manufacturer if a program is troublesome.

The MSD utility

MSD (Microsoft System Diagnostics) is an advanced package which will examine and report on the whole of your system. You can look at the report on screen, as illustrated here, or send the report to a disk file for printing out directly or by using a word processor. There are three file report options depending on how you want the data organized. The straightforward command MSD produces the screen illustrated in Figure 8.2. This consists of a summary, and you can gain additional information on each of the topics shown here by clicking the mouse with the cursor over the topic on which you want more information. The information under the heading of *Computer* will be information on the BIOS chip for any PC clone.

The *Memory* option has been clicked in Figure 8.3, showing a summary of the use of memory in the range from 640 Kbyte to 1024 Kbyte – this is shown in two screens, using the scroll bar at the side to move from one to another, and the illustration shows the upper part. Part of the TSR

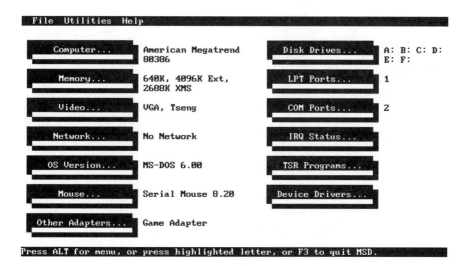

Figure 8.2 *The opening screen of the MSD diagnostic program consists of a main menu.*

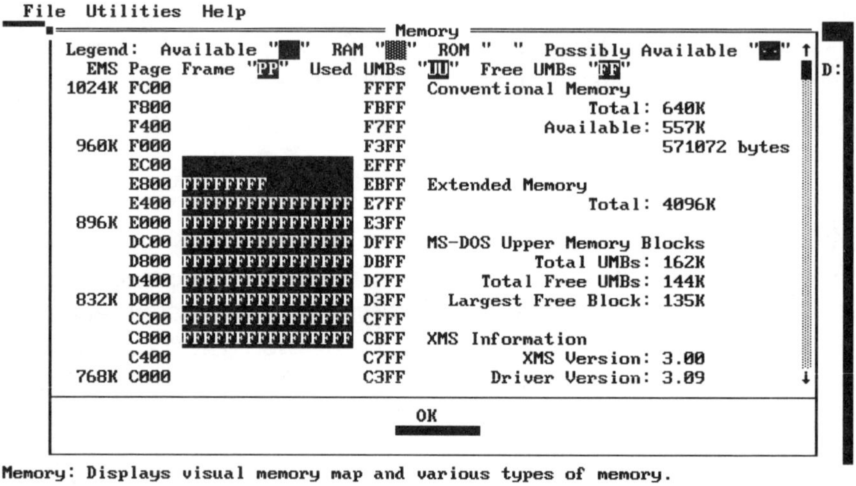

```
 File  Utilities  Help
■=========================== Memory ===========================■
│ Legend:  Available "██"  RAM "██"  ROM " "  Possibly Available "██" ↑
│     EMS Page Frame "PP"  Used UMBs "UU"  Free UMBs "FF"              ▌D:
│ 1024K FC00                  FFFF  Conventional Memory
│       F800                  FBFF            Total: 640K
│       F400                  F7FF        Available: 557K
│  960K F000                  F3FF                     571072 bytes
│       EC00                  EFFF
│       E800 FFFFFFFF         EBFF  Extended Memory
│       E400 FFFFFFFFFFFFFFFF E7FF            Total: 4096K
│  896K E000 FFFFFFFFFFFFFFFF E3FF
│       DC00 FFFFFFFFFFFFFFFF DFFF  MS-DOS Upper Memory Blocks
│       D800 FFFFFFFFFFFFFFFF DBFF        Total UMBs: 162K
│       D400 FFFFFFFFFFFFFFFF D7FF    Total Free UMBs: 144K
│  832K D000 FFFFFFFFFFFFFFFF D3FF    Largest Free Block: 135K
│       CC00 FFFFFFFFFFFFFFFF CFFF
│       C800 FFFFFFFFFFFFFFFF CBFF  XMS Information
│       C400                  C7FF          XMS Version: 3.00
│  768K C000                  C3FF       Driver Version: 3.09          ↓
│
│                            ┌────┐
│                            │ OK │
│                            └────┘
```

Memory: Displays visual memory map and various types of memory.

Figure 8.3 *The MSD Memory option which shows how the main memory of this machine is being used.*

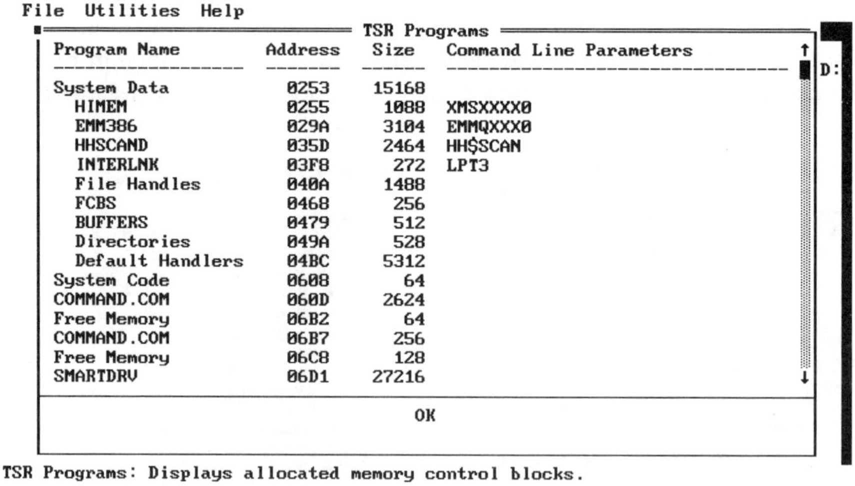

```
 File  Utilities  Help
■======================= TSR Programs =======================■
│ Program Name       Address  Size   Command Line Parameters  ↑
│ ─────────────────  ───────  ─────  ───────────────────────  ▌D:
│ System Data        0253     15168
│   HIMEM            0255      1088  XMSXXXX0
│   EMM386           029A      3104  EMMQXXX0
│   HHSCAND          035D      2464  HH$SCAN
│   INTERLNK         03F8       272  LPT3
│   File Handles     040A      1488
│   FCBS             0468       256
│   BUFFERS          0479       512
│   Directories      049A       528
│   Default Handlers 04BC      5312
│ System Code        0608        64
│ COMMAND.COM        060D      2624
│ Free Memory        06B2        64
│ COMMAND.COM        06B7       256
│ Free Memory        06C8       128
│ SMARTDRV           06D1     27216                            ↓
│
│                          ┌────┐
│                          │ OK │
│                          └────┘
```

TSR Programs: Displays allocated memory control blocks.

Figure 8.4 *Part of the MSD option for viewing TSR locations.*

option is shown in Figure 8.4. This indicates how each TSR and driver uses conventional memory, showing what portions remain in conventional memory even though the main part of each TSR or driver may have been loaded into high memory or UMB.

In general, MSD is more of a programmer's utility, and its most important use is to produce a disk file that can be sent (to Microsoft, for example) for diagnostic purposes. The MSD parameter options, typed following the MSD command (before pressing Enter) are:

/**B**   Produce output in black and white, if your display does not give a clear rendering in colour.

/**F**   Produce a report to a file, and include your name, company and other details, for sending to Microsoft. The filename (such as A:MSDREP.TXT) must follow the /F option.

/**I**   Do not detect machine data – use this only if you have problems in using the straightforward MSD command.

/**P**   Produce a report to a disk file, but with no name or other personal information. The filename must follow the option letter.

/**S**   Produce a summarized report to a specified disk file.

The MSD menus consist of File and Utilities. The File menu contains Find File, allowing you to search the hard disk for any named file. When you use this option you are asked to type the filename, and specify the starting directory, with options to include subdirectories, search the boot drive (usually C:\) or search all drives. Another File option is to print the

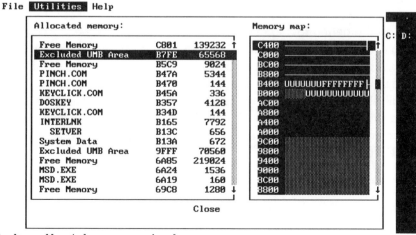

Figure 8.5 *The MSD memory block display for this particular machine.*

```
 File  Utilities  Help
■════════════════════════════ ROM BIOS ════════════════════════════
  F000:B66D System Configuration (C) Copyright 1985-1990, American Megatre  ↑
           nds Inc.,                                                        ▮
  F000:0000 0123AAAAMMMMIIII05/05/91(C)1990 American Megatrends Inc., All
           Rights Reserved
  F000:0050 (C)1990 American Megatrends Inc.,
  F000:0100 ROM BIOS (C)1990 American Megatrends Inc.,
  F000:8000 XXXX88886666----0123AAAAMMMMIIII Date:-05/05/91 (C)1985-1990,
           American Megatrends Inc. All Rights Reserved.
  F000:E0CA R(C)1985-1990,American Megatrends Inc.,All Rights Reserved.,13
           46 Oakbrook Dr.,#120,GA-30093,USA.(404)-263-8181.
  F000:E00E IBM COMPATIBLE IBM IS A TRADEMARK OF INTERNATIONAL BUSINESS MA
           CHINES CORP.
  F000:1A8B If BIOS shadow RAM is disabled,
  F000:2E60 AUTO CONFIGURATION WITH BIOS DEFAULTS
  F000:2FD6 Load BIOS Setup Default Values for Advanced CMOS and Advanced
           CHIPSET Setup
  F000:312E AMI BIOS SETUP UTILITIES
  F000:3273 (ii)  Load BIOS Setup Defaults                                  ↓

                                 OK

Press ALT for menu, or press highlighted letter, or F3 to quit MSD.
```

Figure 8.6 *Using the memory browser, in this example to see the text embedded in a ROM chip.*

report directly, not using a file, and the other main option is to view any of the CONFIG.SYS, AUTOEXEC.BAT or INI files that have been found on the disk.

The Utilities menu consists of Memory Block Display, Memory Browser, Insert Command, Test Printer and Black and White. The memory block display is illustrated in Figure 8.5 above. It contains a list of resident programs on the left-hand side and a pictorial representation of memory use on the right. Moving the cursor to any program name will cause the memory display to alter so as to show the region of memory that is involved. The Memory Browser allows you to search the ROM areas. This is a way of detecting messages and important information, as Figure 8.6 illustrates for a machine with an AMI BIOS chip. You can opt to search for a string of letters or codes if you know what you are looking for. The Insert command allows you to alter some settings in CONFIG.SYS and AUTOEXEC.BAT.

Chapter **9**

The Windows GUI

GUI (Graphics User Interface) is a term that has crept into computing language and which dates from the introduction of the Apple Macintosh. The unique feature of the Mac was (and still is) that the user does not type commands and press the Enter key; instead commands are selected from a menu or a set of pictures (icons) by placing a cursor over the item, using a mouse, and clicking the mouse button. In addition, programs can be independently run within small screen areas, or windows, of their own. In theory, the user of a GUI machine need never type a command, and the Mac makes no provision for any form of type-and-press-Enter commands other than in a 'dialogue box', a small form that the user fills in. The older name for GUI is WIMP, the initials of Windows, Icons, Mouse Programming.

The advantages of GUI methods are that once the basic principles are familiar it should be easier to learn to use any new program. In addition, it is never necessary to exercise your own memory too much – if you want to delete a file, for example, you will see a list of files in the selected directory and you can point to the one you want to erase and drag it to the picture of a waste bin (or press the Del key, which is quicker). Contrast this with typing DEL and then the full path and filename (if you can remember it by then) for each file you want to delete. One of the reasons for the wholesale switch to GUI versions of established programs is to take advantage of easier learning. Other reasons are that the GUI itself contains screen and printer drivers, so that programs using the GUI do not need to supply their own drivers, and because so many programs can benefit from using a graphics screen display that is standardized by the GUI format.

On the other hand, GUI methods are often roundabout and for some actions a lot of mouse movement and clicking is needed. In addition, because a lot of the work of the processor is concerned simply with maintaining the pictorial display, there is less time to spend on the program you want to run. Some database programs do not exist in GUI versions because they would run at an unacceptably slow rate for actions such as searching for data. The PC type of machine allows you to make your own choice, with MS-DOS available for direct commands and for running programs at high speed, with a GUI called Windows available for more leisurely work with the full range of windows, icons and mouse actions. Using Windows does not commit you exclusively to using the mouse for everything, because you can still use the keyboard for actions that are quicker done in that way – every user can devise his/her own best methods of working.

In addition, you can use an intermediate type of GUI in the form of DOSSHELL, which allows you to carry out recognizable MS-DOS commands quickly with window and mouse methods. The choice is entirely yours – you are not locked into any one method, and you can take advantage of new operating systems as they become available. Both Windows and DOSSHELL, however, make use of MS-DOS rather than controlling the computer directly. Later systems will probably break away from this pattern. For more details of using DOSSHELL, and its advantages, see the Newnes *MS-DOS Pocket Book*.

At the time of writing, both Windows 3.1 and Windows 95 were available, but the supply of Windows 3.1 will eventually dry up. If you are using a 386 machine, it is easier to work with Windows 3.1, which is less demanding, but for a 486 or later you should aim to use Windows 95, which is a more capable system. Most of this chapter is devoted to Windows 3.1 because the basics of using Windows are much the same for either version. DOSSHELL is now available only on request.

Windows 3.1

Microsoft Windows 3.1 is a considerably improved version of the original Windows program that Microsoft devised originally as a way of easing MS-DOS users into the type of window-icon-mouse program (WIMP) that could be expected to take over completely when a new operating system finally replaces MS-DOS (which could be many years yet). Unlike the earlier versions, Windows 3.1 is intended primarily for 80386, 80486 and Pentium machines (though it can be used with some limitations on 80286 machines) and it makes better use of the facilities of these more advanced chips.

Windows is designed to run using MS-DOS and to take control over the running of your computer, making it easier to select and run programs. For full details of Windows 3.1, see the *Windows 3/3.1 Pocket Book*, also published by Heinemann Newnes, so that what follows is a summary of the system rather than a full description. To get a reasonably fast performance from Windows you need to be using at least a 386 machine with a clock speed of 20 MHz or more, a large hard disk, and 4 Mbyte (preferably more) of memory. Slower machines, and machines with only 2 Mbyte of memory, can use Windows, but the frustration of waiting for a program to load and run (especially if you remember how easy it was just to type the name and press Enter) will soon drive you back to DOS unless your machine is fast enough to cope.

In addition, you must have a mouse for selecting programs to run and to indicate the way that you want data displayed on the screen. It is possible to use Windows totally without the mouse, but to do so is rather pointless because it would require you to memorize a large number of commands in the form of combinations of keys to press, and if you could do that, why not use DOS directly, or work with batch files? For details of batch files and the simplification they bring to running programs under MS-DOS, see the *MS-DOS Pocket Book*.

The full benefit of using Windows is also apparent only when you use programs that have been designed to run under the full control of Windows. If you have programs such as word processors, spreadsheets, databases and graphics programs that were designed in the days before Windows, then you can, using a 386 or 486 machine, run such programs in a window that occupies only a part of the screen, but you do not have the additional facility of being able to cut and paste data easily (though some cut and paste of text is possible). To gain the full advantage from Windows, then, you will want to use the newer programs that are fully Windows-compatible. Such programs are not compatible with DOS directly – you cannot run any of them by typing the name and pressing the Enter key when the MS-DOS prompt is showing.

You need not worry about your older programs, because it is almost certain that the data files from these programs, if they were industry-standard programs, will be usable by the modern Windows-based programs. Your Lotus 1-2-3 files, for example, can be used by Windows versions of 1-2-3 or by Microsoft's Excel spreadsheet and your database files can be used by programs such as Omnis Quartz or Windows Filer. The superb word processor Word for Windows will read files that are in ASCII code along with some files produced in other code formats by some well-known, established word processors, such as Word for DOS,

WordPerfect for DOS or for Windows and Lotus Ami Pro. Graphics programs are ideally suited for Windows and there are now many that use Windows and can read files produced by older versions, and among DTP programs there are the market leaders Aldus Pagemaker, Ventura Publisher and the newcomers, Serif Page Plus, Express Publisher for Windows and Timeworks for Windows.

Windows preliminaries

Before you start to install Windows 3.1 into your computer, using the set of disks (normally for a 1.44 Mbyte 3½" drive or a 1.2 Mbyte 5¼" drive), you need to know what you are letting yourself in for. To start with, Windows is designed to take control of the running of your computer, making it easier to select and run programs by way of the MS-DOS operating system. Windows is a program which, like any other, makes use of MS-DOS, and in order to take over the running of your computer, some of it must always remain in the 640 Kbyte main memory of your computer. If this memory space is not to be over-crowded, then it is better if your computer does not make use of any other programs that occupy too much of the memory or which would interfere with the action of Windows. These would be desktop utilities such as SideKick or Metro, screen-printing utilities, disk-cache programs other than SMARTDrive, screen-grab utilities like CATCH, SNAPSHOT or PINCH'N'PUNCH. You must also avoid the MS-DOS APPEND, JOIN or SUBST commands.

Windows 3.1 greatly benefits from using the Microsoft disk-cache program called SMARTDRV.EXE (avoid the older version SMARTDRV.SYS), but only if you have an adequate amount of extended memory, 2 Mbyte or more, installed. The version of SMARTDRV.EXE contained in MS-DOS 6.2 is excellent, but if you use the older version in MS-DOS 6.0 you should use it to cache reading only, not disk writing. The APPEND, SUBST and JOIN commands were intended to be used at a time when hard-disk drives were just being introduced on the PC machines, and they should not be in use on any modern machine. In addition, you can usually dispense with any separate MOUSE.COM type of program, because Windows contains its own MOUSE.COM (or equivalent) program for controlling the mouse, and will install it automatically, except for some named mouse types. If you intend to run Windows 3.1 over a network you should *always* start up the network before attempting to start Windows running on any of the networked machines.

Windows does not run fast over a network, and it is often preferable to run local versions.

Windows 3.1 is intended to be used with the MS-DOS operating system, preferably the most recent versions. There may be problems if Windows is used with other operating systems, even those that claim close compatibility. Later versions of Windows are likely to be self-sufficient, containing their own operating system.

Both the CONFIG.SYS and the AUTOEXEC.BAT files will be altered as Windows 3.1 is installed. It is useful to have a printed copy of what these files contain before you install Windows, so that you are aware of what changes have been made. You should also, before installing Windows 3.1, make a copy of each of these files on a spare floppy labelled OLD AUTOEXEC & CONFIG by placing a disk in the A: drive and typing the lines:

```
C:
CD\
COPY CONFIG.SYS A:
COPY AUTOEXEC.BAT A:
```

Wait until you see the C> sign after each COPY command before you type the next one. Each command line is run by pressing the Enter key as usual.

If you are using an old version of MS-DOS, any version number lower than 5.0, it is a great advantage to change over to MS-DOS 6.0 before you install Windows 3.1. MS-DOS 6.0 (or later) allows a much larger portion of the main memory on such a machine to be used for Windows, by placing most of the MS-DOS system in extended memory. It can also place programs (called drivers), which normally sit in the main memory, into memory that is otherwise unused, clearing more useful space in the main memory. The MEMMAKER utility of MS-DOS 6.0 can be run after Windows is installed in order to make the best possible use of the memory that is available.

To get a good performance from Windows 3.1 you need a clock speed of 20 MHz or more (preferably 25 MHz or more), a hard disk, preferably 80 Mbyte or more, with quoted access time of less than 20 ms, and base memory size of 640 Kbyte and as much memory as can be fitted, certainly at least 2 Mbyte and preferably 4 Mbyte or more. In addition, you should have a mouse for selecting programs to run and to indicate the way that you want data displayed on the screen. The full benefit of using Windows is also apparent only when you use programs that have been designed to run under the control of Windows.

Installation

The installation of Windows on to your hard disk is something that you need to plan, because it involves a large number of choices that you need to think about in advance. The Windows 3.1 program suite is delivered on a set of seven 3½" 1.44 Mbyte disks, along with other software, or on the equivalent number of 5¼" 1.2 Mbyte floppies. Your own personal Windows program will contain much less than is packed onto these disks because, in the course of installing Windows, you will make a selection of files from these disks. My own Windows files amount to some 7.3 Mbyte spread over 197 files, the equivalent of just over five of the 1.44 Mbyte distribution disks. This in itself indicates how important is the selection process that is at the heart of Windows installation.

You will be advised by messages on the screen as you install Windows as to what disks you need to insert at any given time. In the descriptions of the main installation method that follows, the use of the standard 1.44 Mbyte 3½" replaceable disks is assumed – disk changes will be required at different points for the 5¼" 1.2 Mbyte disk type.

Installation of Windows 3.1 can be done on a semi-automatic basis, with little or no input from you, because the installation process can sense the details about the machine that formerly had to be supplied by you. The only point to watch is that this automatic sensing has been correct, and when you are asked to confirm it is useful to be able to know whether it is correct or not. This system is called *Express Setup*, and to use it you need know only two things – the type of printer you are using and the port to which it is connected. The alternative is *Custom Setup*, which requires experience and knowledge that you will not have if you are new to MS-DOS and Windows.

For most users, the type of printer will certainly be one of these listed by the Setup program and the few exceptions will be machines which can emulate one of the listed types. If in doubt, use an Epson FX or the IBM Proprinter setting for an unlisted dot-matrix machine, and a Laserjet setting for any unlisted laser printer. If you use a printer which is a PostScript type (of any make) you need to use the PostScript setting. The printer port for the vast majority of users is called LPT1. Some computers may use LPT2 or COM1, and network users will have special designations for the printer that is connected to the network. In general, if your machine is part of a network. Windows should be installed by whoever installed the network.

You can always change your installed options later, for example, to add new printers or other hardware to Windows, and even if you change the screen graphics card you do not need to go all the way through SETUP again in order to ensure that you have the correct files on your hard disk. If you have recently changed to MS-DOS 6.22 from an older version, you should use the CLEANUP command to remove old DOS files (but only once you are certain that you will not need them again). Unless MS-DOS is correctly set up for your computer, you may have problems installing Windows. It is also an advantage to clear and compress the hard disk, using the facilities provided by MS-DOS 6.22. You should certainly do this before you use the DRVSPACE utility (which allows you to store more files on a disk). If you want to use DRVSPACE, you should go for the version included in MS-DOS 6.22 rather than the version in MS-DOS 6.0.

Either type of Setup is started by placing the first Windows 3.1 distribution disk into Drive A and running the SETUP program (type A:\SETUP and press the Enter key). This takes some time to produce any effect on the screen, and it reads in several files which it also lists.

Your first options are to read more about Setup (press F1), to proceed (press Enter), or to leave SETUP (press F3). When you press Enter you then have the choice of Express Setup (press Enter) or Custom Setup (press key C). You can still use F1 for information or F3 to leave Setup, and this applies generally at other stages in the setting up procedure.

When Express Setup is selected, file copying to the hard disk starts almost at once, creating a directory called C:\WINDOWS. You will be prompted by a sound beep and a screen message to change disks at intervals. These intervals are not necessarily equal, they depend on how much of each disk needs to be read. While files are being copied, you will see a percentage completion display on the screen, and similar displays will also be used to show the progress of installing other sets of files.

During the time when Disk 4 is in use, the screen will brighten to its Windows format and you may be asked to re-insert Disk 3, with choices now being made by using the mouse arrow and three boxes. Move the arrow (by moving the mouse) to the box marked Continue, and press the left-hand mouse button to continue. From this point on, accessory programs will be placed on the hard disk, and you will asked to re-insert Disk 4 followed by the others.

In the course of using Disk 5 you will be asked to select from the large list of printers. You should select the printer or printers that you intend to use. For a printer that is not listed, select a compatible model (see your printer manual for a list of compatible models). When you have selected a printer or printers, you are asked to select the printer port, which will be LPT1 for most users. You will then be asked to insert Disk 6 so that the printer(s) can be configured. There is also a disk of International Printer Drivers you may be asked to use if your printer is one that is not on Disk 6. The printers list includes many obsolete models and also many printers which have never been available in the UK.

The final actions are to run through a set of Windows actions and search the disk for DOS programs that can be run using Windows. You are offered a chance to run a short tutorial on the use of Windows at the end of this process – do this if you have never used Windows before. Finally, you are asked to remove all floppy disks and reboot the computer by selecting the Reboot option button with the mouse and pressing the left-hand mouse button.

Once installation has been carried out and the machine has been rebooted, typing WIN (press Enter) will start Windows with its Microsoft flag logo display which is eventually replaced by the working screen of Windows 3.1.

Opening your Windows

After all this, having done a lot for your installation of Windows, you are ready for what Windows can do for you. You have seen already, as you made use of the latter parts of the Setup program, how Windows carries out its actions, placing information into window-shaped parts of the screen, and allowing you to make selections by putting the mouse-pointer onto a box and clicking the mouse button (always the left-hand button for Windows 3.1 use). This is only one visible part of the action, however, and the whole resources of Windows are not so easy to grasp as all that. You know now, however, how to select using the mouse, which is such an important part of the Windows action.

Even with suitable hardware you might not be getting the best perform-ance from Windows. You must make quite certain that you do not have your memory cluttered with programs that reside in the memory and which are no longer needed. You also need to use programs that are Windows-compatible. You can run almost any program using Windows, but there are many old-established non-Windows programs like WordStar, Lotus 1-2-3 and dBase-3 which on a 386 machine with inadequate memory can use only the full-sized screen. This bars you from using a smaller window so that you can see two or more such programs, or programs along with Windows accessories, running side by side.

This problem can be dealt with by changing to programs that have been written for use under Windows, like WordStar for Windows, Lotus 1-2-3 for Windows or the spreadsheet Excel, or by changing to the later 80386 or 80486 type of machine with adequate memory. Your old data files that were created by the earlier program can still be used if you change programs, because the modern Windows-compatible programs will be able to read the data files that were made by pre-Windows programs.

Having installed Windows on a suitable machine, you are ready for what Windows can do for you. Windows allows you to carry out the actions of MS-DOS, plus many enhancements, in a pictorial way using the mouse. A program or action is represented by a small image, known as an icon. By double-clicking on a program icon, for example, you can run that program. There are so many menus in Windows, and so many ways of carrying out each action, that a lot of time can be wasted in exploration. Some built-in accessories of Windows will be used here to illustrate the actions that you need to learn in order to use Windows effectively with your own programs.

The actions of click, double-click and drag, and the use of scroll bars, are important for using Windows. Clicking means pressing down and releasing the left-hand mouse button – as the name implies this should be a rapid action, not a slow press and release. Double-clicking means that you carry out the click action twice in quick succession – speed is important because the machine will not recognize a double-click that is done too slowly (it will assume that you have single-clicked twice). You can alter the rate that the machine will accept; see the *Windows 3/3.1 Pocket Book* for details.

Dragging is a way of moving whole windows (to shift their position) or edges of windows (to change their size). To drag, you press down the left-hand mouse button and move the mouse with the left-hand button held down. You release the button when you have finished with the move action. The scroll bars are bars that appear at the side (and sometimes also at the bottom) of a window. By dragging a small box in the scroll

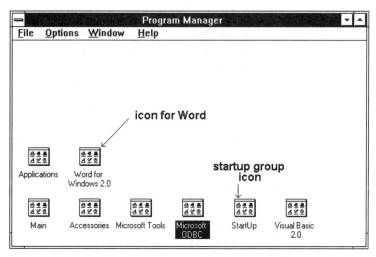

Figure 9.1 *The Program Manager of Windows 3.1, showing the icons.*

bar you can make the screen scroll up or down, allowing you to see an image that is larger than the size of the window.

Windows is normally started, after the computer has been switched on, by typing WIN and pressing the Enter key (this sort of action is written as WIN <Enter>). When Windows 3.1 starts work the screen view is of the management section, called Program Manager, Figure 9.1. If this screen view is not visible but there is an icon labelled as Program Manager at the foot of the screen then double-click on this icon. If there is another window (usually labelled Main) lying over Program Manager then double-click on its Control Menu bar (see Figure 9.2).

The Program Manager consists of a background or Desktop screen (a pattern or wallpaper display) with a title bar, four main items, and a set of icons at the bottom of the screen. The icons are used to fetch other programs (referred to as applications), as an alternative to the use of the File and Windows items in the menu. The mouse methods of program management are more useful for all but the advanced actions, and at this stage in Windows, they require more attention. The first set of mouse actions concerns its use to pick groups, making use of the group icons. When you start to use a Windows program it will be represented by an icon in one of these groups, preferably the Windows Applications group.

We can illustrate the use of a group with the Accessories group. To make use of the Accessories group place the pointer on the Accessories icon and double click. This will produce the separate icons for the group,

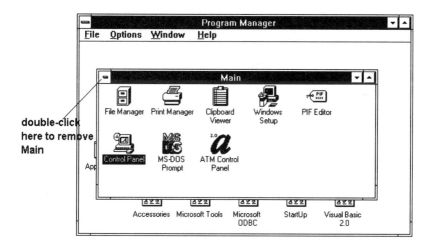

Figure 9.2 *Removing the Main window if it lies over the Program Manager.*

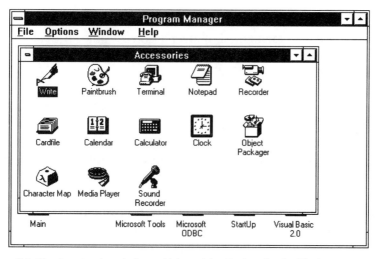

Figure 9.3 *The Accessories window, which contains the icon for the Clock.*

as illustrated in Figure 9.3. Select the Clock program in this group, by double-clicking on its icon. The result is shown in Figure 9.4. The clockface appears on the screen, partly covering the Program Manager and Accessories displays. In other words, the clock appears in a window that has been pulled down over both of these other displays, obeying the Windows rule that the most recent window always appears over the earlier windows that are still being displayed.

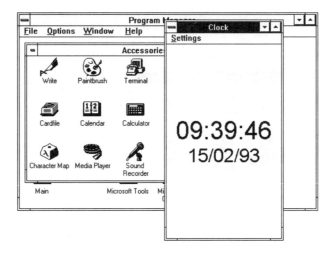

Figure 9.4 *The Clock, opened by double-clicking on the icon in the Accessories group.*

The clock is fully operational, showing correct time in digital form, assuming that your computer has had its time correctly maintained. Now the position of the clock is probably far from ideal, so that it can be useful to be able to move this window. In fact, you can move it, alter its size, or tuck it away for future use: all important actions for programs that are designed to run under Windows.

To move the clock put the pointer on the black bar at the top, near the name Clock, avoiding the symbols at either end of the bar. The pointer should be a single arrow. Press the mouse button down and drag the clock to another position, Figure 9.5, so that a ghosted outline can be seen to move across the screen, following the mouse pointer. Now release the mouse button, and the clock will appear in its new position, Figure 9.6, still keeping perfect time. The digital display is a default, and you can alter this to a two-hands display if you like, by double-clicking on the Settings menu and clicking on Analog in place of Digital – this setting will remain in use until you alter it again.

To resize the clock place the arrow on any part of the frame round the clock window so that the usual single arrow change to a double-headed arrow. Press the mouse button and move the mouse to change the size of the clock frame. Figure 9.7 shows this at a point when the size has been changed, but before the mouse button is released. If the pointer is placed at a corner of the frame, the two-way arrows face diagonally,

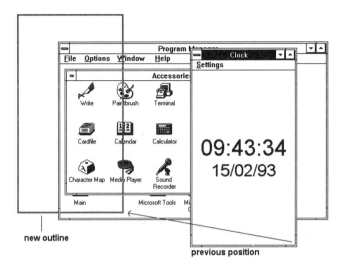

new outline

previous position

Figure 9.5 *Dragging the clock window to a new position – the ghosted outline allows you to see where it will lie when you release the mouse button.*

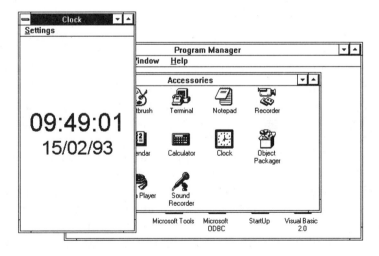

Figure 9.6 *The clock released in its new position.*

allowing both height and width of the frame to be expanded or contracted as the mouse is moved with the button pressed.

The right-hand side of the title bar contains two arrow symbols. The up arrow, if selected, will make the clockface occupy the whole screen, covering the Program Manager display. Select this arrow and click on it

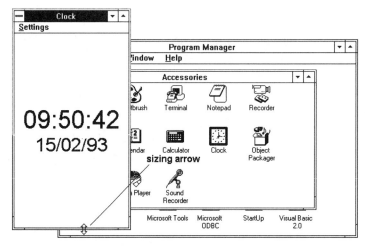

Figure 9.7 *Altering the size of the Clock window – the two-headed arrow indicates that size and not position is being altered.*

to maximize the clock display. The other arrowhead (down-facing) will minimize the clock, shrinking it back to the inactive icon in the Accessories window. When a box with two arrowheads appears (in a maximized view), clicking on this has the same effect as Restore, restoring the former size of the window. If you double-click on the Control Menu bar at the left-hand side of the title line, the program is closed. Unless the data was saved, it will be lost, so that a reminder will appear if any data was being used. Minimizing a program to an icon does not close it, and its data is left in place to appear when the program is restored.

Minimizing is a useful way to put a program temporarily out of the way, but keeping its data and settings. It is quicker than resizing with the mouse. Windows therefore provides a visual way of carrying out file actions and running programs, allowing you, when using Windows programs, to operate inside these window spaces that can be moved around the screen and resized. Using the 80386 or later machines, separate non-Windows programs can be run simultaneously inside different windows, providing true multi-tasking, limited only by the amount of memory and your ability to see the contents of each window. Windows 3.1 also has the ability to manage sound systems (either multimedia sound and vision displays or synthesizers) if the necessary hardware is present; see Chapter 11.

The Clipboard

The Windows Clipboard is a facility that allows data to be transferred from one Windows program to another. The action is basically similar to the Cut or Copy and Paste action in a word processor, but Clipboard allows both text and graphics to be transferred, retaining the format of the original. It also allows a Paste Special action to be used, in which you can specify the format – for example to copy text only or text along with formatting codes. The material in the Clipboard is usually held only temporarily, but can be saved as a Clipboard file for use at another time. Previously saved Clipboard files can also be loaded back into the Clipboard for use in any other program.

The material that has been clipped can be pasted into any program that is capable of accepting it. This is the main limitation of Clipboard, so that a graphics image cannot be pasted into a word processor of a type incapable of handling such images (not that there are many such word processors nowadays). The distinction between bitmapped and vector graphics images is also important – an image produced by a CAD program cannot normally be pasted into Windows Paint, for example. The Clipboard is most effective when both programs are specially-written Windows programs (programs designed to run under Windows) rather than older programs which are simply being run by Windows. The power that Clipboard provides, however, is very attractive, particularly when you bear in mind the contortions that are needed in order to cut material from one program and transfer it to another when using only DOS. By taking advantage of the Windows facility to keep several programs running together, the Clipboard offers fast transfers, offsetting the (short) time that is spent when switching from one program to another. The use of the Clipboard is confined to text when earlier programs (not designed to run under Windows) are being used.

The object linking and embedding facilities of Windows 3.1 are more powerful than the simple copying action of Clipboard. As a standard example, suppose you have a graphics program which has just produced a file called LOGO.PCX, a small logo image for your notepaper, and you are also using a word processor that accepts PCX files so as to allow a graphics image to be put on a page. The word processor is currently working on a file called MYHEAD.DOC. The LOGO.PCX image can now be embedded in MYHEAD.DOC or linked to it.

If embedding is used, the logo image appears in the MYHEAD document, and if you want to alter the image, you simply double-click on it. This will start the graphics program running so that you can use it for editing the image, and when you close the graphics program you will be back editing the MYHEAD document with the image changed. The image is saved along with the document. Linking looks similar, and you can double-click on the image to edit it just as if you had embedded it, but the image file is not saved as part of the document, and when you are working with the graphics program, any changes that you make to LOGO.PCX will appear in the version that is linked to MYHEAD. This happens whether MYHEAD is being edited or not at the time when the LOGO file is being changed. If the same image is used by a several other documents, all of them will change when the linked image is changed.

Programs that are designed to run only under Windows are invariably much larger than the older generation of DOS programs. In part, this is because they do so much more, but whether what they do is essential to you is another matter. Because Windows allows the use of additional (extended) memory (above the normal 640 Kbyte that is assigned for a DOS program), lack of memory need not be a problem for programs running under Windows – but the fact remains that there are some older programs that can work perfectly satisfactorily in 64 Kbyte, putting today's bloated monsters of 1 Mbyte or more to shame. In addition, there is the speed handicap. A program running under Windows is invariably running slower than it could run under MS-DOS, though comparisons are difficult to make.

What it boils down to is very much a matter of taste and since Windows is optional you are not forced to use it, unlike the GUI operating system of some other machines. Windows 3.1 has been a notable success, with users turning over to Windows programs in millions, and this is certain to continue because few users ever want to return totally to direct MS-DOS methods. The next version of Windows is likely to be a complete operating system in its own right rather than a bolt-on addition to MS-DOS, and its performance will be enhanced accordingly.

Windows 95

Windows 95 is the preferred option for any fast modern machine. The minimum specification for satisfactory use is:

- A fast 486 or Pentium (or later) processor working at 66 MHz or more

- RAM memory of 8 Mbyte or more, preferably 12 Mbyte

- A hard drive of 350 Mbyte or more, preferably 850 Mbyte

The installation of Windows 95 will make allowances for any machine with less than 7 Mbyte, installing a modified version, but this is rather less satisfactory, and you would be as well served by using the last version of the older Windows, V3.11.

Windows 95 does not use the Program Manager approach that was used in Windows 3.1, and when you start Windows 95 you will see a toolbar appear at the bottom of the screen. This has on its left-hand side a 'button' marked Start, and this is how programs are launched. When you place the mouse pointer over this button and click the mouse button, you will see a menu appear, and when you move the pointer to an item on this menu, other menus will appear with no need to click again – this is a much faster procedure than the old Windows. An example of the fully-expanded set of menus is illustrated in Figure 9.8.

The program names and icons that appear in these Start menus can be run by a single click, not the double-click that is used for the older system. Once a program is running, its name appears on the Toolbar strip

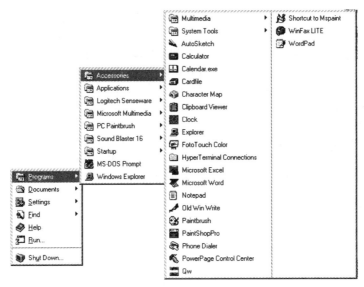

Figure 9.8 *A typical set of Start menus, fully opened out.*

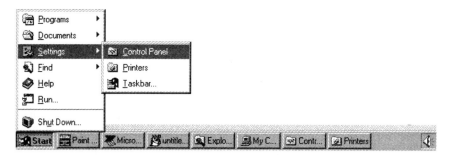

Figure 9.9 *The Toolbar, showing several programs running, and how to start Control Panel.*

(which can be moved to any of the four edges of the window). The Toolbar is illustrated in Figure 9.9, along with the pointer settings for the Control Panel.

The old File Manager display of Windows 3.1 has been replaced by the Explorer, which uses a single split window, allowing you to drag files from the file display on one side to any directory (now called folder) on the left. The main theme of Windows 95 is to work using the mouse rather than typing, so that you will see fewer boxes appearing that require an entry from you. At the same time, the use of longer filenames (more than 8 characters, up to 255) allows you to type filenames that you can recognize easily, although older programs will use only an abbreviated version of such filenames.

Many actions, such as renaming a file, have been made much easier, though in a different way. For example, if you want to rename a single file you simply click on the filename and then edit it, but there is no provision for altering a set of filenames it's quicker to go back to DOS for such actions!

For full details of Windows 95, see the *Windows 95 Pocket Book* (Butterworth-Heinemann), and if you are familiar with Windows 3.1, take a look at *Moving Up to Windows 95* (Sigma Press).

Section IV

Everybody's problems

Chapter **10**

Printers

A printer is by now such an essential part of a computer system that you tend to forget that it is seldom part of a computer package. Unlike the monitor, the printer is practically always bought separately, so that your choice of printer is important, since it may outlive several computers (in the sense that the computers are replaced by newer devices but the printer is not). In addition, it is reasonable nowadays for all but the smallest system to use more than one printer or to share the use of a printer among several computers.

Output on paper is referred to as hard copy, and this hard copy is essential if the computer is to be of any use in business applications. For word-processing uses, it's not enough just to have a printer, you need a printer with a high-quality output whose characters are as clear as those of a first-class electric typewriter. For desktop publishing you will need a laser or inkjet type of printer. Given that the use of a printer is a priority for the serious computer user, what sort of printers are available? The answer is any type that comes with a parallel interface, which means virtually any printer currently on sale, though some bargain offers may have the alternative serial type of interface. The parallel type of interface (see Chapter 5), means that each bit of a byte will be signalled to the printer over a separate wire, so that when control wires are also added, the cable between the computer and the printer contains a large number of wires, and is usually of the ribbon type.

The system is simple and easy to set up, you simply plug in the cable and start printing. This is due to the standardization of the connections by the printer manufacturer Centronics, so that the parallel system is often referred to as a Centronics Interface Parallel port, or the Centronics parallel interface. The main problems of parallel data transfer as far as printers are concerned are of line length and the rate at which signals are sent. Each signal is a pulse, a very short, sharp change of electrical voltage, and the longer the length of the cable, the more likely it is that signals will be corrupted. This can happen in two ways. One is that the short sharp changes of voltage are rounded out into long smooth changes, ruining the timing of the signals. The other problem is that a signal on one line will be picked up on another line, causing an error at the printer end of the cable.

The practical effect is to restrict parallel printer leads from computers to around 2 metres, and most printer cables are only 1 metre long. You can, if you need to, get around this restriction by using repeaters, amplifiers which restore the correct signals at the end of a long line, but few PC owners take this way out of a cable length problem. Serial printers require more care over the connecting cable (of fewer wires), and infinitely more fuss and bother over setting up of software. The only advantage of using a serial printer is that the cable can be longer than the usual 2 metre length specified for parallel printers. Only a few portable machines nowadays omit a Centronics port, and it's best to avoid such machines unless the price is irresistible. Printers that are used with small computers will use one of the mechanisms that are listed in Figure 10.1.

Of these, the impact dot-matrix type is still the most common. A dot-matrix printer creates each character out of a set of dots, and when you look at the print closely, you can see the dot structure. Most of the dot-matrix printers are impact types. This means what it says, that the paper is marked by the impact of a needle on an inked ribbon which hits

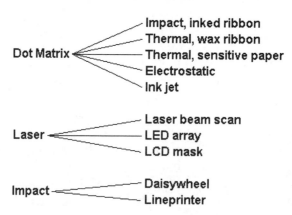

Figure 10.1 *Printer mechanisms – not all are commonly used nowadays. The most common modern types are dot-matrix, laser, ink-jet and (particularly for colour printing) the wax-transfer form of dot-matrix.*

the paper. There are also thermal and electrostatic dot-matrix printers. These use needles, but the needles do not move. Instead the needles are used to affect a special type of paper, and since these types of printers are hardly ever used for PC applications we need not devote any space to them. The only exception is the wax ribbon thermal type such as the IBM Quietwriter which can produce print of excellent quality with very low noise levels, but such printers are not common so far, though they are now being produced by several other manufacturers as a method of producing high-quality colour printing.

Impact dot-matrix

The older type of dot-matrix printer used a printhead that contained 9 wires or needles in a vertical line. This 9-pin, or 9-wire, printer type is still manufactured in large quantities, and some are sold at very high prices because of their particularly robust construction or high-speed printing or both, but the trend nowadays is to 24-pin printers. By using two slightly staggered vertical rows of 12 pins each, these printers can print at a high speed and with excellent quality with none of the dotty appearance that has been associated with dot-matrix printers in the past. The noise level of such printers is usually higher than that of the 9-pin types, and ribbon life is shorter since so many more pins are striking the ribbon. Most 24-pin printers have built-in fonts so that a range of different type styles can be obtained by operating switches on the printer itself, or under software control from your word processor if (and only if) it has a suitable printer driver. Each pin is, incidentally, of a smaller diameter than a human hair.

There is a huge range of manufacturers, but most printers are set up so as to emulate either the IBM range of Proprinters or the Epson types – most 24-pin printers will provide emulation of the Epson LQ type. Some other emulations are not always satisfactory, because they can lead to a 24-pin printer being used to emulate a 9-pin type, so that the superior quality that the 24-pin type can provide for graphics use is not being utilized. In general, the best text quality is obtained from the built-in fonts of such printers so that it is an advantage to use a word processor that allows the printer's own fonts to be used rather than forcing the printer to draw the font patterns imposed by the software.

Another advantage of using the built-in fonts is that they print very much more quickly. Most dot-matrix printers can be fitted with optional extras that considerably extend the actions of the machine. Memory is the most common type of add-on card, permitting the use of a printer buffer of, typically, 16 Kbyte or more to be used. This allows the computer to dump text into the buffer for printing, so that the computer itself is free for other uses and is not tied up in the slow business of printing. Printers which come with a built-in 16 Kbyte buffer usually provide for this to be considerably extended by means of plug-in cards.

Another provision, mainly on modern 24-pin printers, is for fontcards, containing ROM for fonts that are additional to those built into the printer. The standard parallel interface can be supplemented by a serial interface, using a plug-in card, various colours of ribbons (sometimes also the option of carbon film ribbon) can be used, and most printers can be supplied with a single-sheet feeder to replace the standard tractor or friction-fed mechanism.

Laser printers

The ultimate in print quality at the present time can be provided by the laser type of printer, which also includes variants such as LED bar printers and LCD-mask printers. These are fast and silent in action. The laser types are page printers, meaning that it is necessary to store a complete page of information in the memory of the printer in order to print the page. Fonts can be built-in, added by way of a cartridge (surprisingly expensive), or transmitted in bitmap form (downloaded) from the computer. When elaborate graphics or downloaded fonts are used, this can require a large amount of memory – some laser printers require 2 Mbyte or more of memory to function satisfactorily. Some LED bar types are not page printers, and can work line by line, requiring very little memory. All quoted printing speeds for printers of any kind tend to be optimistic. Laser printers work on a principle called Xerography (Trade Mark of the Xerox Corporation) which was invented in the 1960s. The similarities between the laser printer and the Xerox photocopier are so close that the two products can be made in one assembly line. A page cannot be printed until the drum which is used to store the 'bit-image' of a page is fully 'printed' with electrical charges (the drum is usually 'printed' more than once to form a page, but the printing does not start until all the print-bits are assembled in the memory). In addition, the mechanism depends on the paper being moved continually against a drum, rather than in one-line steps.

When elaborate fonts are used, this can require a large amount of memory – laser printers require at least 2 Mbyte of memory to function satisfactorily for DTP work which contains large fonts and graphics. This does not mean that you cannot use a laser printer with only 512 Kbyte of memory – such a printer can cope perfectly well with ordinary pages of book printing that contain no graphics. It is also possible to handle graphics if you confine the resolution to 150 dots per inch or lower. In addition to memory, the laser printer also contains the components of a computer, with a main processor of its own. This processor is used to convert the pattern of bits in the memory into instructions for guiding the laser beam and turning the beam on or off so that the drum can be discharged in the correct places. This is how the laser printers can work with such a wide range of fonts and sizes and also with graphics. Note, however, that a laser printer cannot reproduce true half-tones, because each dot that it prints is black. Half-tones can be simulated by mixing black dots and white spaces, but this leads to a coarse appearance on a 300 dots per inch printer and is really satisfactory only on a typesetting machine which works at 1200 to 2400 dots per inch.

The laser printer uses a drum of material which is electrically charged (by an electrical discharge or corona through air which as a by-product produces ozone). Ozone is not good for you, it is a hazardous gas, and the laser printer should be used in a well-ventilated space. Any electrically-charged object will attract small particles to it, and the purpose of charging the drum is to make it possible for finely-powdered ink (called toner) to adhere to the drum. The material of the drum, in addition to being a material which can be electrically charged, is also photoconductive, meaning that it becomes an electrical conductor when it is struck by light. When the drum becomes conductive, the electrical charge will leak away so that the drum can no longer attract particles. The principle of the copier is to make the drum conductive in selected parts, and this happens when the material is rendered conductive by being struck by a light beam. The laser beam, being a beam of concentrated light, has the effect of making the material of the drum conductive where the beam strikes it at full brilliance. The beam intensity (on or off) and direction is controlled by the pattern of signals held in the memory of the printer, and enough memory must be present to store information for a complete page.

This requires about 0.5 Mbyte as a minimum for text work, and 2 Mbyte or more if elaborate high-resolution graphics patterns have to be printed.

As the drum rotates, the laser beam, under the control of the built-in computer, is scanned across the drum. Once this scanning process, which uses the same principles as mechanical television in the 1920s, is complete, the drum will contain on its surface an electrical voltage 'image' corresponding exactly to the pattern that exists in the memory, matching the pattern of black dots that will make up the image. Finely-powdered resin, the toner, will now be coated over the drum and will stick to it only where the electric charge is large – at each black dot of the original page.

The coating process is done by using another roller, the developing cylinder, which is in contact with the toner powder, a form of dry ink. The toner is a light dry powder which is a non-conductor and also magnetic (some machines use a separate magnetic developer powder), and the developing cylinder is magnetized to ensure that it will be coated with toner as it revolves in contact with the toner from the cartridge. A scraper blade ensures that the coating is even. As the developing cylinder rolls close to the main drum, toner will be attracted across where the drum is electrically charged – relying on the electrical attraction being stronger than the magnetic attraction. Note that two forms of attraction, electro-static and magnetic, are being used here.

Rolling a sheet of paper over the drum will now pass the toner to the paper, using another corona discharge to attract the toner particles to the paper by placing a positive charge on to the paper. After the toner has been transferred, the charge on the paper has to be neutralized to prevent the paper from remaining wrapped round the drum, and this is done by the static-eliminator blade. This leaves the toner only very faintly adhering to the paper, and it needs to be fixed permanently into place by passing the paper between hot rollers which melt the toner into the paper, giving the glossy appearance that is the mark of a good laser printer. The drum is then cleared of any residual toner by a sweeping blade, re-charged and made ready for the next page. Figure 10.2 shows the principles in a diagram.

The main consumables of this process are the toner and the drum. The toner for most modern copiers is contained in a replaceable cartridge, avoiding the need to decant this very fine powder from one container to another. The resin is comparatively harmless, but all fine powders are a risk to the lungs and also a risk of explosion. The drum was, in early printers, coated with selenium, which is not a material that should be handled if you can avoid it, and which will give off very toxic gases if it is ignited (selenium is a close relative of sulphur, and is flammable). Never

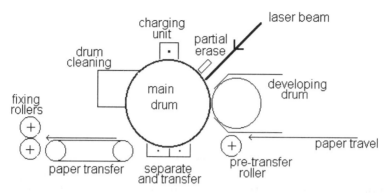

Figure 10.2 *A typical laser printer mechanism. The fixing rollers are heated so that paper emerges hot. Sticky labels can be affected by this heating and only laser-printer grade labels should be used.*

open a laser printer which is operating. Most printers now use the less objectionable zinc oxide form of coating, and several are now using organic photoconductors (OPC) which are of low toxicity and which do not have to be returned to the manufacturers after replacement.

Drum replacement will, on average, be needed after each 80,000 copies, and less major maintenance after every 20,000 copies. Some models use a separate developer powder (a magnetic powder) in addition to toner, and the developer will have to be replenished at some time when the toner is also exhausted. The Hewlett-Packard Laserjet machines use a cartridge which contains both the photoconductive drum and the toner in one package, avoiding the need for separate renewal – the life is quoted at about 3500 sides at average print density of word-processed text, but this figure will be drastically reduced if you print a lot of dense graphics and fonts. Long-life cartridges are available.

There are some types of printers which are classed as laser printers but which do not use a laser beam. These are LED-bar or LCD-mask types which use the same principles of light beams affecting a charged drum, but without the use of a laser beam scanning over the drum. These types are not page printers, and can work line by line, requiring very little memory. They were originally intended as replacements for daisywheel and dot-matrix printers for word-processing applications rather than as a competitor for the DTP type of laser printer, but several have now been developed into suitable machines for DTP use, and some of them now make use of PostScript.

The quoted speed of most laser printers, 4 pages per minute upwards, refers to printing repeated copies of a single page and does not refer to normal printing, which is always very much slower. This is because a substantial amount of the time that is needed for printing consists of building up the instructions in the memory for forming the charge pattern on the drum, and if this pattern remains fixed, page printing can be as fast as the speed of the drum permits. When each page is different, and in particular if graphics images are used on each page, the time needed to form the pattern makes the printing rate very much slower. All quoted printing speeds for printers of any kind tend to be highly optimistic. The fastest printing is of pages of text that use one of the built-in fonts of the printer.

Laser choice

Most of the lower-cost laser printers use as their standard the Hewlett-Packard Laserjet, and emulate the codes used by that printer (which is itself not expensive). This type of printer is excellent for word-processing and graphics, and virtually all word-processing and graphics software will provide printer driver software for the H-P printer. The most recent Series 4 Laserjets are excellent machines which rival the PostScript form of machine for serious use. The more expensive option is the use of a printer that is fitted with the PostScript command language (Trade Mark of Adobe Corporation) and with enough memory to allow Post-Script to be used. This automatically pushes up the price of the printer to considerably higher than the level of the H-P type of machine, though these prices have been dropping steadily since the first laser printers were introduced, and are still falling at the time when this is being written.

The advantage of using PostScript is that it allows for graphics to be scaled without loss of resolution, and it also allows a large variety of fonts to be used in soft form. Several makes of printers (notably H-P and Texas) allow PostScript to be added in cartridge form so that the low-cost printer can later be upgraded to full PostScript capability. This is a useful path at a time when the price of the PostScript addition has been falling rapidly. The use of Windows TrueType soft fonts and modern graphics methods now make PostScript less attractive for the user who does not require compatibility with high-cost machines. In addition, the PCL form of printer control used for the later (Series 3 and 4) Hewlett-Packard printers is now being used to an increasing extent.

A soft font is one that is loaded into a word processor or DTP package as a file, as distinct to the use of a plug-in cartridge to the printer, and the considerable advantage of a PostScript soft font (and of any TrueType font) is that it can be scaled. The amount of scaling that can be used in practice depends on the amount of memory that is available at the printer, and when work is scaled, text and graphics will remain in registration – something that is very difficult to achieve when bit-mapped fonts are used. The most important advantage of using PostScript for a laser printer is that PostScript is also used by professional typesetting machines, so that the laser printer can be used to produce draft copies of material from PostScript files that can also be used by Linotronic or similar typesetting machines for very high quality output. As a comparison, laser printers are generally capable of 300 dots per inch, and this can be doubled by altering the electronics inside the printer. A typesetter, by contrast works with 1200 – 2400 dots per inch, allowing good half-tone reproduction which is not achieved with the lower resolution of the laser printer.

At the time of writing, the most attractively-priced laser printers were the Windows printers, which made use of Windows to store image patterns and deliver them to the printer ready to use, rather than have this processing done within the printer itself. This saves the cost of a processor and memory within the printer, and if you are using Windows 3.1 these printers are worth considering. If you intend to upgrade to Windows 95 you need to be more careful, because there is a chance that some printers of this type might not work. Unless you plan to use only Windows 3.1 (not DOS, not Windows 95) for all the life of the printer, wait until you can be sure that the printer you like is compatible with Windows 95.

Using the laser printer

The working heart of the laser printers is known as the engine and there are only a few basic engines (such as the Canon) used in all of the laser printers that are currently manufactured. This makes it all the more surprising that there is not more interchangeability between makes for such items as font cartridges, toner cartridges, replacement drums and so on. If you are buying a laser printer for the first time, it pays to enquire on the costs of these consumable items, because these costs are often much more important than the cost of the printer itself.

Paper is the most consumed item, and laser printers use, as might be expected, the photocopier grade of paper whose cost is at least twice that of ordinary paper of the same density unless you shop around. The reason for the additional cost is the way that the toner is deposited on to the paper. The paper should consist of fibres which are all aligned along the longer axis of the paper, making the paper behave more uniformly when subject to electric charges (and discharges). It also allows the paper to feed through the machine with less tendency to curl. In addition, since the toner is fixed to the paper by fairly intense heating, the paper must not darken or curl when it is heated. These requirements make the paper more expensive to produce, though some shopping around can reveal better prices than can be obtained from local suppliers. Try Viking Direct if you cannot obtain low-cost paper locally – and look at their catalogue also when you want other consumables like toner.

Whatever is claimed by manufacturers, the use of very heavy (more than 70 grams per square metre) and expensively finished paper is not justifiable. Such paper will often feed badly, forming ridges, and will allow toner to smear. Very heavy paper will stick in the printer or cause loud protests from the rollers. Lighter and more absorbent papers usually produce better results – try cheap grades first and always try a sample before you buy several hundred packs.

The other major costs are replacement of toner and of the print drum. Toner is a fine powder which must not be allowed to spill into the atmosphere, and the print drum is constructed using a photosensitive material which must not be handled or exposed to light. In addition, if the element selenium is used for the print drum there is a hazard due to the poisonous nature of selenium (and the selenium dioxide which is generated if it overheats). Some manufacturers have made the replacement of both toner and drum particularly easy for the user, for other machines the task is far from easy and better done by a maintenance mechanic. Maintenance does not simply cover the replacement of the toner and drum, it also concerns cleaning. Because of the way that electric charges attract all small particles, laser printers tend to become clogged up with fine dust, composed of stray toner and house dust in almost equal measure. Dust is, as always, an enemy of mechanical parts, so that cleaning and lubrication schedules are of considerable importance.

Lubrication almost always uses silicone oils – mineral oils are totally forbidden on the plastics which are almost universally used for bearings on light machinery. Users are often advised to start a new run of a major printing with a fresh toner supply. Though it is inadvisable to start a run when the toner is almost finished, replenishing toner is not advisable before a major piece of work. When toner has been replenished, the first set of pages may be over-inked and badly smudged. Following toner replenishment, always make some test copies onto absorbent paper until you are sure that the toner is flowing correctly – I have never experienced these problems with the Laserjet machine. Note that toner cannot be vacuum-cleaned – it is too fine to be retained in the bag of the cleaner unless you are using a Medivac™ or Nilfisk™ type of machine. If you get toner on clothes, wipe it off with a cloth moistened in cold water. Avoid hot water at all costs – it can melt the toner into the fabric, making it impossible to remove.

The laser printer is not the universal answer to printing requirements, because though the cost of buying a laser printer has dropped dramatically since the early days, the price of maintaining such a printer is still high. The consumables are costly, particularly toner (powdered ink), and when more extensive servicing is required the chemicals that are involved are toxic and expensive. Laser printers, incidentally, should, like photocopiers, be used only in a well-ventilated space.

Ink-jet machines

The ink-jet printer, which operates by squirting tiny jets of ink at paper from a set of miniature syringes, is a close second in quality to the laser printer. The bubble-jet technology, developed by Canon and Hewlett-Packard, has been widely adopted to make printers of remarkable quality and reliability at comparatively low prices. This technology has been said to originate in the observation that a hot soldering iron laid on a hypodermic needle caused a drop of liquid to be ejected. The principle is to use a head consisting of fine jets (of a diameter narrower than a human hair) each provided with a miniature heater wire. Passing current through the heater for a jet will expel a tiny drop of ink, so that by driving these heater wires with the same form of signals as a dot-matrix impact printer, the ink can be deposited in the same character patterns.

A more recent development is the piezo-ink jet printer developed by Epson. The principle here is that part of the jet path is through a piezoelectric crystal (one that deforms when a voltage is applied to it), and when a voltage pulse is applied to the crystal it contracts, forcing ink from the jet. In principle, this can be a faster mechanism because there is no need to wait for the heating or cooling of an element, and it is possible that finer jets could be made. The technology is still new, but the first machine to become available, the Stylus 800, is very competitive in price and looks as if it will be more economical in ink cartridge prices, since the jets are said to be less prone to clogging.

If you need graphics printing with large areas of pure black, you might find an ink-jet superior to a laser in this respect. The paper quality is less critical, but the cost of ink can be high because the whole set of nozzles is usually renewed along with the ink reservoir. Several firms offer recycled cartridges at a considerably lower price than the original, and there are also re-inking kits available.

The ink-jet types are line printers and are remarkably silent, most are considerably quieter than laser types (many of which have a noisy cooling fan). The speed of printing is not as high as that of the slowest laser type, but for many applications this is of little importance, and the ink-jet types have the advantage that they can also print in colour. The colour ink-jet printers are sold at prices that are not substantially higher than the cost of black and white, though the cost of consumables is much higher.

Daisywheel printers

An older technology is represented by the daisywheel printer. This uses a typewriter approach, with the letters and other characters placed on stalks round a wheel. The principle is that the wheel spins to get the letter that you want at the top, and then a small hammer hits the back of the letter, pressing it against the ribbon and on to the paper. Because this is exactly the same way as a typewriter produces text, the quality of print is high, but the machines are very noisy and slow. They have been almost entirely replaced by 24-pin dot-matrix and laser types for most business applications, though they can still be found (and heard) in some smaller offices. Their lack of versatility has been the deciding factor in their decline because changing a font has to be done by changing the print-wheel, and no graphics can be printed. The other side of all this is that daisywheel printers can now be bought for much less than the cost of a portable typewriter. Unless you need only text and can buy very cheaply, avoid this type of printer.

Plotters

The other way of producing paper output, mainly for graphics, is the pen plotter, which uses a set of up to 8 miniature ball pens moving over a sheet of paper. The top-quality plotters (such as the Hewlett-Packard range) are expensive, but ideally suited to CAD work. Lower-cost plotters are available, but mostly for work on small paper sizes compared to the A3 which is a standard for graphics work. Plotters are ideally suited to producing output in colour because each pen can be of a different colour.

The pound sign

The English pound sign still causes more problems that any other aspect of printing, even now. The root of the problems is that the ASCII code that is used for each character was of US origin, where the English pound sign is not used, and, before the PC machine became the standard, various codes were used. The PC machines allocate code 156 for the English pound sign and on an English keyboard this sign is on the SHIFT-3 key. Pressing this key will make the pound sign appear on the screen, but only if the CONFIG.SYS file contains the line that uses KEYB UK to establish the UK keyboard correctly, and the COUNTRY line correctly set up also.

When you have constructed or adapted your own PC you may find that you have bought a US keyboard, so that the KEYB UK instruction must not appear in the CONFIG.SYS line. On such a keyboard you can make the sign appear by using the combination Alt-156. Hold the Alt key down, and type the number 156 on the numeric keypad at the right of the main key set, then release the Alt key. This does not work if you type the number on the keys at the top of the main keyboard.

Being able to get the pound sign on the screen is not the same as getting it on a printer, because what a printer produces for the code 156 depends on what is stored in its own memory. Leaving aside older devices such as daisywheel printers (on which you have to be certain that the £ sign is on the daisywheel before you start to worry about which code it uses), the use of dot-matrix and laser printers ensures that you can print any character – if you know how. If you are using the IBM Proprinter, the pound sign requires the same code of 156 as the PC screen, so that there is no problem.

If you are not using a Proprinter, but some other impact dot-matrix printer, then it is likely that it can be set to emulate the Proprinter. This is usually done by setting a tiny switch (a DIP-switch, Figure 10.3) inside the printer, and setting this for Proprinter emulation ensures that what you see on screen can be printed.

ON or 1

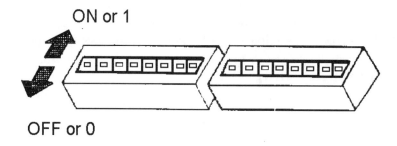

OFF or 0

Figure 10.3 *A typical DIP-switch set inside the casing of a dot-matrix printer.*

If your impact dot-matrix printer cannot be set for Proprinter emulation you might have more of a battle on your hands. The older Epson impact dot-matrix printers used code 163 for the pound sign when they were using the US character set, and the code 35 when switched to the English character set. Code 35 is normally the hashmark (#, sometimes referred to confusingly in US manuals as a pound) and it would have been more logical if the sign for the dollar had been used.

If you are using a printer that follows this scheme then you need to set the switches inside the printer to the English character set and to use the hashmark in place of the pound sign – the setting can also be done using software signals. Some software, notably Lotus 1-2-3 will co-operate with this by allowing you to choose a symbol for monetary units for the printer, so that a £ on screen is converted to a hashmark when printed.

Modern Epson machines allow you to use more than one character set, and by selecting the one that uses 156 for the pound, usually termed the IBM #2 set, your troubles are over. Laser printers almost universally use the Hewlett-Packard standards, and by using the PC-8 character set of this machine you can ensure that the £ sign will be printed for code 156. The PC-8 set is selected when you set up the printer, using its internal menu system – consult the manual for your printer to see how to select a character set.

Printer problems

Mechanical problems with printers are surprisingly rare, despite the complex mechanism of many printers. My own record is one jammed sheet and one jammed ribbon in the last ten years. One problem that can arise at times is paper jamming. Ordinary paper seldom causes any difficulties but envelopes and multi-part stationery can be troublesome, and some printers are very much better than others at handling such awkward materials. Particular care is needed in handling sticky labels. These are normally mounted on a backing sheet, but if they become detached they can deposit their adhesive on rollers within the printer, causing problems until the rollers are removed and cleaned. If you want to use self-adhesive labels with a laser printer, make certain that the labels are laser-grade which can withstand the heat of the laser printer.

A genuine mechanical fault will usually require the printer to be returned to the manufacturer or to a service agent. Before doing this, try the printer's self-test and test output routines. If a printer can produce a perfect copy of its test page (usually a sample of fonts) then a mechanical or electrical fault is unlikely to be the cause of the problem, which is more likely to arise from incorrect settings either at the printer or at the software. Remember, for example, that when you use the Windows printer manager the printer will not necessarily start printing immediately – it may wait until a queue forms or a 'print now' instruction is issued.

The majority of printer problems concern failure to print what is expected, whether it is the £ sign on accounts statements or the box shapes on a screen display. The usual reason is, as noted earlier, that the printer has not been configured to use the correct character set. This will be PC-8 for laser printers, and the IBM #2 set for dot-matrix printers (most ink-jet printers follow the Laserjet sets). Attending to this point solves some 95% of these problems. In other respects, printers are remarkably reliable.

If the main ON indicator fails to appear this can usually be traced to the mains switch of the printer switch being off or the power cable not correctly plugged in. Failing to print anything when the printer is clearly switched on is usually due to the printer being off-line or the data cable incorrectly inserted – the data cable connector should be firmly inserted and locked in place. Software can also cause what looks like printer errors, for example when the software is set for a line width that is greater than the page limits.

One important aspect of software concerns printer drivers. Most printers will perform adequately with text, but can sometimes provide very disappointing results with any kind of graphics work. This is often due to a poor printer driver, the software that matches the printer to the computer. Some Proprinter drivers, for example, when used with a printer that is set for Proprinter emulation, can produce very poor results, and it is always difficult to find out where the blame lies. A few printer manufacturers will provide drivers for Windows and for each of the most popular non-Windows software packages, but if you have to use the driver that comes with the software, particularly if your printer is set for an emulation, you may encounter driver problems. Sometimes an up-grade is available, but if your printer is of an obscure make (and you do not always know what is an obscure make from a US point of view) your only hope is that the printer manufacturer can provide a better set of drivers. Users of Windows who use only Windows programs are by far the most fortunate in this respect because they need only a single printer driver for Windows.

Chapter **11**

Multimedia and other connections

In this chapter we shall be looking first at adding multimedia to your computer, and the rest of the chapter concerns other hardware items that you might consider adding. These latter items are not essentials, but many of them considerably extend the use of the PC into fields that are well beyond the usual office software. One such addition, the modem, has already been covered in Chapter 5, mainly because its use is so tied up with the use of the serial port that the two must be described together.

CD-ROM

Multimedia requires a sound card and a CD-ROM drive, and the need for CD-ROM at least is highlighted now by the wide availability of software on CD-ROM. Windows 95, for example, is distributed in this way as a popular option, and at a lower price than the disk set. Other software is rapidly following suit, particularly DTP and graphics software which can use the added space on the CD to put in items such as clip-art, which would otherwise occupy an inordinate number of floppies. The amount of data that is involved in some modern programs, and in items such as graphics files, makes the use of conventional magnetic disks inadequate – a set of 100 3½" disks is a major expense in media, let alone in contents.

Figure 11.1 shows the shape and dimensions of the low-cost Mitsumi CRMC-LU005S drive, which fits into a standard half-height 5¼" bay. When a CD is to be inserted, an inner tray slides out to allow the CD to be placed on its spindle. When this inner shelf is re-inserted, the front panel (Figure 11.2) reveals a drive-on light, a volume control and a headphone jack. This allows the use of the drive for playing ordinary audio compact discs (or you could listen to your data). The output audio signal is typically 0.6V rms at 1 kHz. This is a typical single-speed drive,

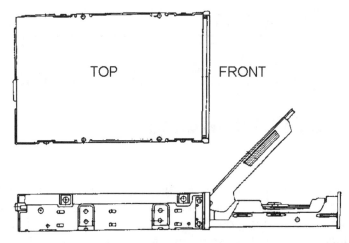

Figure 11.1 *The Mitsumi CD-ROM, showing the top view (closed) and the side view (open).*

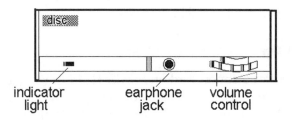

Figure 11.2 *The CD-ROM player front panel, with its indicator, headphone jack and volume control that allows music CDs to be played back.*

and is no longer supplied, because the double-speed drive is now a standard. The older type of single-speed drives are very well suited to the older and slower computers, however, and can often be picked up at very low prices (typically £75 or so).

CD-ROM can be used in two ways that are described as Mode 1 and Mode 2, both of which are supported by the Mitsumi drives. In Mode 1, storage capacity for the conventional 12 cm disc is 540 Mbyte per disc and the transfer rate for the drive is 150 Kbyte per second. In Mode 2, the capacity is 630 Mbyte and the transfer rate is 175 Kbyte per second. Typical access times for a block of information range from 0.25 seconds to 1.2 seconds, depending on where the block is located.

The unit uses three connectors: the power connector (as for a floppy or a hard drive) feeding +5V and +12V DC, an audio connector for amplifiers, and the data interface. The audio interface uses a 4-pin connector with connections:

1	Right channel
2	Earth
3	Left channel
4	Earth

The data interface uses a 40-pin insulation displacement connector. On this connector, all the even-numbered pins from 4 onwards are earth connections, and the other pins are connected as follows:

1	Address bit 0 input
2	Address bit 1 input
5,7,9,11	Not connected
13	Interrupt output
15	Data request output
17	Data acknowledgement input
19	Read enable input
21	Write enable input
23	Bus enable input
25	Data bit 0
27	Data bit 1
29	Data bit 2
31	Data bit 3
33	Data bit 4
35	Data bit 5
37	Data bit 6
39	Data bit 7

Mitsumi later launched a faster (double-speed) drive, the CRMC-FX001D whose specification includes transfer rates of 300 Kbyte per second for Mode 1 and 350 Kbyte per second for Mode 2, with typical access times of 0.25 to 0.39 seconds. There is a 32 Kbyte buffer memory built in, and the mechanism uses dust sealing to eliminate a past problem of CD-ROM drives, corruption of data because of dust on the optical pick-up. A later model now offers quad-speed performance. This and the Panasonic drive type CR526B are very often supplied as part of a multimedia upgrade kit.

It is important to buy a drive of the type described as *multi-session*, because such a drive opens the way to the use of the Kodak CD-ROM system. The basis of the Kodak system is that you buy a blank disk, paying about £5, and send your photographs to be processed into digital form onto the disk at a cost of around 35p per frame. If you have a large number of photographs to transfer you may fill the CD, but most users will not, and the unused space is wasted unless your CD-ROM system permits multi-session use. This allows the disk to be sent more than once to have new images added (not replacing the earlier ones), and the point about specifying a multi-session drive is that a single-session drive can gain access only to the original set of images, not the ones that have been added. Using a multi-session CD-ROM system allows you to add images until the disc is full, and to display and use these images as you wish.

The standard single rate was never really useful, and the double-speed drive became a standard even in 1992. You can now buy quad-speed (x4) and hex-speed (x6) drives, and this trend is likely to continue so that the quad-speed will become the plain vanilla offering and the higher speeds the exotic options. Note also that though you can use a sound card with CD-ROM under DOS, virtually all of the CD-ROM material that you will want to use will need Windows, so that there is a software requirement as well as a hardware requirement. For some specialized purposes (for example, some music composition programs) you might be able to use DOS, but this is becoming more unusual as the months go by.

In addition, new units will be *Plug'n'play* compatible. Plug'n'play means that the BIOS chip on the motherboard of the computer is capable of interrogating chips in devices such as CD-ROM to find out how they need to be set up, and the operating system does the rest. Note that a combination of software and hardware is needed. If your computer has an old BIOS chip, or if your new hardware isn't to Plug'n'play standards, or if you are not using Windows 95, then you can forget it, and you will have to install the hardware in the old-fashioned way.

If you intend now to upgrade to CD-ROM and a sound card, you should look for Plug'n'play units and buy the fastest CD-ROM drive you can afford. A computer that uses an EIDE card should be able to plug straight into the CD-ROM drive with no need for any interface card; simply using the hard-drive connector cable. This leaves the sound card independent, and if this also is a Plug'n'play design, you have very little to do except to plug the card into place. Plug'n'play is particularly advantageous for a sound card, because the installation of a sound card was always an awkward exercise, involving jumpers and lines in the CONFIG.SYS file. With Plug'n'play and Windows 95 you need never be aware that you have a CONFIG.SYS file, and hardware installation is quite unbelievably easy.

Multimedia uses

Multimedia means that a CD can contain program code, digitized images and digital sound, and all of these can be used by the computer if a suitable sound system board is also added. This allows you to use encyclopaedia CDs that display text, images and can also play sound. This is a very far cry from the feeble sounds that most people associate with computer games, and for anyone whose interests include music the use of multimedia promises effects that are unmatched by any other medium, even videotape, because of the extent of the control that the computer program exerts over the display. For current multimedia work, the double-speed drive is adequate, a quad-speed is very desirable (particularly for video animations) and hex-speed will soon become affordable.

The CD-ROM drive is installed in the same way as a floppy drive is installed and, as usual, it helps if your casing has a generous number of drive bays. The CD-ROM drive is usually a half-height 5¼" unit, so that there should be no problems in fitting it to the standard bay. You have to be certain that your interfacing is correct. Some models will connect only to a specified sound board, some use a small interface card that fits in an expansion slot, and the more recent units will connect to a hard-drive connector, particularly if the EIDE type of hard drive interface is used.

Unless your CD-ROM unit is of the modern Plug'n'play variety (and the rest of the computer is able to use it), you will have to install software to make use of the CD-ROM drive. This varies considerably from one unit to another, but the example of a Panasonic drive connected to a Soundblaster card is typical, and also widely available, so that a description of this process is useful. If you are working under Windows 95 you should use the Control Panel option for installing hardware.

Software installation

Make sure that all cables are plugged into their correct places and the lid is shut; switch on the power. You should hear the high-pitched whine of the hard drive motor start and settle to its final speed and the computer should boot up normally. Before you start installing software, you can check that the turntable for the disk will eject. Press the eject button – and keep your hand out of the way as the tray pops out. Pressing the tray gently in, or pressing the eject button again, will make it wind in all the way. This checks that the power supply to the CD-ROM is working, and once the software is installed all should be well.

The software consists of short drivers and will be supplied on a 3½" disk – for the example of the Panasonic CR562 CD-ROM the software was on a single 720 Kbyte disk. Note that some users have reported difficulties that are resolved only when the sound card is installed ahead of the CD-ROM drive, so that you might prefer to carry out the software installation in that order. The CD-ROM software disk will usually contain a README.TXT file and you should print this out as soon as possible – *preferably* before you install the software. As usual, this installation disk is placed in the A: drive, and if you are using Windows 3.1 you should exit back to MS-DOS. If you are using Windows 95 Control Panel, follow the instructions that you see on screen. The following describes installation using MS-DOS.

Type A: \ (press Enter) to switch to the A: drive and (for this example) type INSTALL (and press Enter). Some software may require you to type SETUP rather than INSTALL – check with the leaflets that are packed with the drive. What follows now is the description of the procedure for the Panasonic drive coupled to the SoundBlaster16 sound card. Other packages will be very similar to this, and there will be significant differences only if you are working with different interfaces.

The software installation for the Panasonic drive starts with a menu of five items, and most users can opt for the defaults that are already filled in. You are asked for the drive from which installation is being carried out, with a default of A – alter this only if you happen to be installing from any other drive, which is unlikely. The *target drive* is shown as C: (the hard drive) and, once again, you are unlikely to want to alter this. The target directory is shown as PANA\ and you can accept this default also unless you particularly need to use another name.

The *sound selection* item default is Panasonic AT Interface, and you need to alter this if you are using another interface, such as the SoundBlaster Pro. The last item is IO Port selection, which is shown with a default of 220H. This is a reference number for input and output, and you should use this default unless you know that it cannot be used because of other equipment. If your machine contains only a disk drive card and a video card it is most unlikely that anything else is using this address, but if you have added cards for anything else, such as a scanner, check with the manual for the other equipment to find what IO address is used, if any. Only if there is a conflict do you need to alter the address for the CD-ROM to the only alternative of 240H. You must not alter this number unless the setting on the CD-ROM drive has also been altered. These settings are made automatically in a Plug'n'play device.

Once the menu has been completed the files are copied to the hard drive. Another set of choices then follows. You are asked to select an interrupt number ranging from 0 (60H) to 7 (67H), with a default of 0. Use the default of 60H unless you know that this interrupt number is used by other equipment.

On the Panasonic installation you are then asked to use the program MKECDAPL, which has been placed on your hard disk. The menu screen disappears, and you are returned to MS-DOS, with the C:\PANA directory selected. Note that if you use MS-DOS 6.0 or higher, the MSCDEX.EXE file will also be contained in your MSDOS (or DOS) directory. This is the program that allows MS-DOS and Windows to work with CD-ROM and the sound board. All of this should place a line such as:

```
C:\PANA\MSCDEX.EXE /D:MSCD000
```

in the AUTOEXEC.BAT file. Check that this line is placed correctly – one installation program placed this line at the end of the AUTOEXEC.BAT file, following a line that started Windows, so that it had no effect until Windows ended. When the line is correctly in place, booting up into MS-DOS will show a message such as:

```
CD ROM Device Driver Version 4.03
Copyright (C) Matsushita-Kotobuki Electronics Industries Ltd,
1990,1991,1992,1993. All rights reserved
Device Driver Name = MSCD000
Supporting the following units :
    unit 0  id 0 MATSUSHITA CD-ROM CR-562-x  0.76
1 CD-ROM drive(s) connected
CD-ROM device driver installed

MSCDEX Version 2.22
Copyright (C) Microsoft Corp. 1986-1993 All rights reserved.
    Drive J: = DRIVER MSCD000 unit 0
```

This message will be seen briefly as Windows 95 starts, so that you can check that it exists (but don't expect to have time to read it!) The really valuable information here is that the CD-ROM will be referred to as drive J in this example. If you are not using the letter D: for any other drive, this is the one that will be assigned for the CD-ROM.

If you use Windows 3.1, you can use the Windows sound routines to play audio CDs. Windows 95 goes one better – simply insert an audio CD into the drive and it will play from start to finish.

You can use multimedia CDs even without the sound card installed, though obviously there will be no sound – the point is that the multimedia programs do not pack up if there is no sound card. If your interests do not demand sound, then you can use the text and picture facilities from CD-ROM without the need for a sound card and loudspeakers. Apart from anything else, this saves space on your desk.

Each CD-ROM that you use will need to be installed, usually under Windows. The installation will create a directory on your hard drive which contains essential routines that the computer needs to be able to get at quickly, using the codes on the CD itself as a backing store. This means that you need adequate hard drive space if you intend to use several CD-ROM discs, particularly encyclopaedias. This point is stressed again later in this chapter.

Sound boards

The PC was intended as a machine for business use from the time the PC/AT was launched, and though a small loudspeaker is built in, this is intended only for delivering warning notes. For many users to date, this is all that is needed, and there are still many users who resent even a warning bleep for a computer. If greater capabilities for sound output or input are required, this must be added to the basic PC machine by way of inserting a sound board into a spare slot. Another path is to use add-on devices that plug into the parallel port – these are made 'leadthrough' so that the printer can still be connected. The use of a card in a slot is by far the better solution, and nowadays it would be short-sighted to specify the older type of 8-bit sound card. A modern 16-bit sound card need not be expensive, and you should consider the Orchid Sound Producer Pro or the SoundBlaster Pro-16 before making any decisions. It make sense to install both sound card and CD-ROM drive as one package.

The most common reason for wanting to add a sound board to a PC is so as to be able to use multimedia fully. This is particularly important if your interests are in music or other sound-related topics. The other common business requirement is to be able to input sound, such as a spoken commentary on an image or a document. The software for such requirements is built into Windows 3.1 and Windows 95, but the hardware has to be added. Multimedia packages usually consist of CD-ROM player (often the Panasonic model) with sound card (usually SoundBlaster), loudspeakers, microphone and a range of software.

The Windows software includes a number of files which use the WAV extension, and which will provide sound outputs such as chimes and ringing notes. These can be set so that they are associated with events such as incoming messages, errors, disk saving and so on, acting as a reminder of what you are doing. You either like this or loathe it, and it is up to you whether you activate it or not. By contrast, the Sound Recorder software in Windows (Versions 3.1 and 95) allows you to input sound from a microphone or a cassette recorder and record it as a WAV file (which will be large even for comparatively short bursts of sound). You can then embed this into a document (for example) in the form of a visible icon. When you are reading the document, clicking on the icon will play back the sound recording. There are also rather crude sound editor facilities (echo, mix, blend, play faster or slower) built into this software.

The Media Player software of Windows allows devices such as audio CDs or video discs to be controlled by the computer, and a standard form of musical instrument digital interface (MIDI) allows the computer to be used to control electronic instruments such as synthesizers. One point to remember, however, is that digital sound requires large amounts of storage space on a disk – one often-quoted example is that one minute of speech can require more than 600 Kbyte of disk space. This requirement can be reduced by using data-compression techniques built into the software, but you need to be aware of how much disk space is likely to be used.

A typical sound-board package will consist of the card itself, a stereo amplifier, two loudspeakers (not always of high quality) and a microphone, probably with a MIDI interface built into the board, and additional software which can be used to replace or supplement the software supplied in Windows 3.1. The board is plugged into a slot in the usual way, and the other units connected to it.

In selecting an add-on sound system, you should consider the quality of sound from the loudspeakers before anything else, because you may find it difficult to live with poor-quality sound. The lower-cost packages are likely to omit a microphone and, more important, will omit software that allows sound files to be compressed so as to take up less disk space. The software is the next important item, and should be suitable for your requirements. One standard item should allow you to add narration to documents or images, and another piece of software that is often included is a voice synthesizer which will read text from a spreadsheet or a word-processed document (usually contained in the Windows clipboard). This latter application can often be, on its own, justification for adding a sound board. Some boards come only with software suitable for games.

When you want to add a multimedia item, such as the Microsoft Encarta encyclopaedia, you follow the normal pattern of installing software. Each CD that you buy will require some software to be installed on the hard drive and, following installation, when you want to use the package you need to insert the CD and run the software from the hard drive.

For example, if your CD-ROM drive is D:, you would use the Windows Run command to install the software from the CD. When the RUN window appears, you would type D:\SETUP and start the process running. For Windows 95, you use the Control Panel Install Software option, and follow the instructions (though you can also use the RUN option from the Start menu). This will ensure that the multimedia software is ready to use for future occasions.

For further details of multimedia use, including image manipulation, see *Multimedia on the PC* (PC Publishing, 1994).

Scanners, digital cameras and graphics pads

A scanner is a useful method of obtaining a graphics file from a picture, so that you can edit, select and transform the picture to your own needs. Scanners come as low-cost hand scanners (Figure 11.3), medium cost roller types, or high-cost A4 types, and unless you need large images or text scanning, the hand type will suffice. If, on the other hand, you need to do a lot of detailed work on printed images, some of which are more than 4" wide, or you need to scan text into image form and then convert this to text form, only the roller or page scanner will suffice, and you are committed to spending something in the order of £300 - £600 at least. These up-market scanners come packaged with OCR (Optical Character Recognition) software that will convert the image of words (which is just a pattern of dots) into computer text (using one code for each character).

Fitting a scanner often involves the use of an interface card, and you will need to set jumpers on this card in accordance with the manufacturer's instructions unless the device is to modern Plug'n'play standards. The option is the type that plugs into a parallel port and, as this requires no settings at all, it is very convenient. If you are worried about having

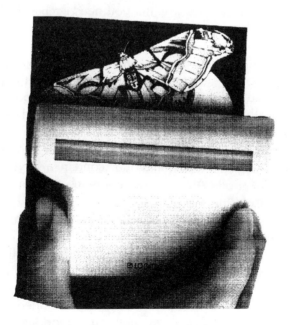

Figure 11.3 *Pulling a hand scanner over an image, courtesy of Logitech.*

to share the printer port with a scanner, then install a second printer port – parallel-port cards are cheap and easy to install. You need to consider in advance of buying a scanner whether you want to use it predominantly for monochrome or colour images. Older scanners were used under DOS, and some performed poorly under Windows, but the more modern types are intended for use under Windows only.

All scanners require fairly elaborate software. At its simplest, the software will be of the bit-mapped graphics variety, so that the scanned image is converted into a file. This is usually of the 'standard' TIFF (Tag Image File Format) – the trouble is that different manufacturers have their own ideas about TIFF standards so that software that reads TIFF files might not read *your* TIFF file. The purchase of a scanner is therefore something that requires some planning and thought. Another consideration is time and memory, because it can take five minutes or so for a flat-bed scanner to read an A4 page, and some 2 Mbyte of disk space to store the file. Some makes, however, come with a very useful package of software. The Logitech PowerPage, for example, (a roller type) can be used for images or text reading, as a fax scanner (using a modem) or as a copier (using the printer).

You need to ask yourself whether you need such a device. If you are working with DTP and need an occasional graphic, the extensive libraries of clip-art will probably hold enough to keep you in images for the foreseeable future, even if you sometimes need to have an image converted for special purposes. The second point is the type of image you want to read. If this is a simple black-and-white logo or pattern, then it is quite likely that a low-cost hand-held scanner could provide as much as you need, particularly if you can retouch the image later by any type of drawing program. If on the other hand you frequently need to be able to read photographs or drawings that contain a large range of grey shades, then you will certainly require a much more elaborate flat-bed scanner, with matching software.

To read text, you must scan the text as a graphics image and then use OCR (Optical Character Recognition) software that will recognize the shapes of letters and convert into a file of ASCII codes that *can* be read by a word processor. Do not, however, expect better than 99% accuracy in reading, so that your text will require editing unless the OCR software is particularly good. This applies to full-page scanners, because using OCR with a hand scanner is too slow and laborious (not to mention error-prone) to consider. Even a flat-bed scanner is laborious to use unless the OCR software can control the scanner and convert the scan directly into text, preferably quickly. Do not expect miracles in this respect – a time of ten seconds for an A4 page is considered fast. If the software does not control the scanner directly, you will have to scan to a file and then set the OCR software to work on the file, which is all more time-consuming.

Prices of scanners and their associated software have dropped fairly rapidly, and at the same time the capabilities of simple hand-held scanners (particularly if you can use them on a good flat drawing board) are improving. The roller type of scanner, of which the Logitech Power-Page is typical, is remarkably capable, and its OCR software is among the best available. Most scanners allow digitization of photographs by means of 'dither' in which shades or colours are converted into dot patterns of differing density, sometimes with a switch-selection of dot size. This allows reproduction of almost newspaper standard for some images.

A more recent type of image-grabber uses a miniature video camera as the scanning device, because the price of video cameras has now dropped below the level of the traditional flat-bed scanner. The video camera does not have to be of broadcast standards, or even of home-video standards, because it will be used in a well-lit studio situation. In this way, there is no necessity to work from photographic intermediate images, because you can scan a live scene. It is also easy to scan material on paper, making this the most versatile of image input systems. Another recent option makes use of the Kodak CD-ROM system along with a CD-ROM reader in the computer, allowing photographed images to be digitized in the Kodak workshops and read from the CD-ROM disk – this requires the Kodak software, or an image package that can work with the Kodak images.

Prices at the time of writing vary over a wide range. A small hand scanner with resolution of 100 - 400 lines per inch and a 16-level grey scale can be obtained at around the £75 to £110 mark. This includes software that will work with files of the IMG, PCC, PCX, MSP and TIF varieties, so allowing you to use almost any of the popular graphics packages for editing your images if the supplied software is not adequate. If your have an incompatible graphics program, you may find it possible to convert file formats – the Public Domain Software Library lists image file conversion software and one graphics file editor among its very useful offerings.

The PowerPage scanner is listed at £299 at the time of writing and unless you must use a flat-bed scanner this provides for all the scanner actions you are likely to want. Though the action resembles that of a fax reader, the base of the scanner can be removed so that the scanner will pull itself along a page which is too large to be inserted into the usual slot. The software is unusually comprehensive, and the accuracy of OCR makes this a very useful machine.

Going up to the flat-bed type of scanner, prices have taken a tumble recently and the Hewlett Packard Scanjet 3P is now on offer at £229, with the more capable model 3C at £609. At the other end of the scale, you can spend £10,000 on a flat-bed A4 machine that will cope with any sort of printed material, and what you spend depends, as always, on what you *must* have.

The use of a scanner pre-supposes that you already have an image on paper, but a more radical alternative is a video still camera that is directly connected to the computer or one which stores its image on an internal disk. For example, the £99 Connectix Quickcam connects to the parallel port and can produce an image of up to 320 x 240 pixels using 64 shades of grey. It can also produce moving images at up to 12 frames per second, which is suitable for slow movement. This allows you to copy photographs and drawings as you would with a scanner, but also to photograph objects directly.

Other digital cameras from Logitech and Canon are much more expensive and use an internal disk so that they can take photographs independently of the computer. The digital image files from these cameras can be transferred from the disk inside the camera to your PC for further processing, and since many graphics packages can convert a photographic image into a line drawing this is a superb way of creating artwork for illustrative purposes. The Fotoman and the other digital camera, the ION PC from Canon, are not cheap, but their versatility makes them worth considering as a way of producing image files. One point to watch is that no supplementary lenses are currently available for the Fotoman, making it difficult to work in close-up, though with patience this can be achieved. If you are accustomed to using a modern 35 mm SLR camera you may find the digital cameras rather difficult by comparison.

Digitizers

Another option is the use of graphics tablets or digitizers, which require an interface board. The graphics tablet looks like a small drawing-board with a stylus, and its action is to control the cursor. The action is not like that of a mouse, because each position on the graphics board corresponds exactly to a position on the screen, allowing you to trace drawings, for example. Pressing the stylus corresponds to clicking the left-hand mouse button, and the driver software allows for a key combination to carry out the action of the right-hand mouse button. You can use this to create drawings either directly or by tracing an existing image, and at the current price of equipment this can often be a cheaper option than a scanner. The Genius range of graphics tablets is by far the most comprehensive. Most graphics tables allow you to control a CAD program such as AutoSketch directly from the tablet, and most allow tracing of photographs or drawings. Do not, however, rush into buying a tablet, even at an attractive price, unless you are sure that it will operate for all the software that currently uses the mouse. You will probably need separate drivers for DOS and for Windows and possibly others for software such as AutoSketch CAD. All of this can make the use of a tablet much less attractive, particularly if you have to plug the mouse back in and re-arrange the software so as to make use of some programs.

Fax and fast modems

The current fashion for the Internet has made the possession of a fast modem an important part of a modern computer system. My own view of Internet proceedings is that it rather resembles the worst of CB radio, but this does not invalidate the usefulness of a fast modem, because you can use such a modem for sending and receiving fax. Unlike Internet, you do not have to shell out £10 per month or more to be able to send and receive fax, and fax messages seem to have the desired effect of being read by the people they are intended for. In addition, there are lots of people who have no computers but who do use a fax machine.

Fast fax modems are obtainable as cards or as stand-alone units, and though the stand-alone units are more expensive and need a power supply they are often easier to get. The card is an attractive answer because it need only be plugged into the expansion slot and to the telephone socket and set up (and Plug'n'play will soon be along). On the debit side, it cannot be shifted from one computer to another and, more seriously, most cannot be switched off independently of the computer. When you can switch off your modem, you make it impossible for anyone to gain access to your computer, and in these days of hacker scares that's an important point.

A lot of fax effort is used simply to transmit typed documents, and since typed documents nowadays are usually prepared on a word processor it makes more sense to do the transmission from the computer file rather than from the paper. This cuts out one form of distortion and it also leads to clearer received copy, because you can receive a fax and print it on plain paper rather than on the thermal paper that low-cost fax machines use.

As always, you need a combination of hardware and software. The modem is the hardware, and a fax modem will come with software such as WinFax Lite that allows you to use the fax facilities. Windows 95 contains Microsoft Fax, and you can decide for yourself whether to use the software that comes with the modem or the software that is built into Windows 95. One point to watch is that if you allow the WinFax software to install itself so that it starts automatically with Windows, you will not be able to use the modem with applications like the Telephone Dialer of Windows 95, the Cardfile of Windows 3.1 and 95, or the Hyperterminal of Windows 95 unless you close the WinFax software first.

The facilities that can be offered by the fax software may include call scheduling, meaning that calls can be made in the cheap (or, more truthfully, less expensive) telephone times, an ordinary telephone socket can be used (no need for a dedicated line) and the software can store telephone numbers for your fax contacts so that the whole use of the system can be done under computer control. Since a fax is received as a disk file, this can be edited by graphics programs, using the usual techniques of zooming, scale alteration, inversion, mirroring and so on.

Typical prices for fax modems now start at just over £100 for the excellent Swedish Intertex modems that can use the 14,400 bits per second speed that is now regarded as a minimum for fast communications.

Teletext cards

Teletext cards are intended to take the signals from a TV aerial and extract the text (and graphics) information from the normal BBC and ITV transmissions (*not* from satellite transmissions). These provide a wide range of information from the Ceefax and Oracle services, and because there is no form of charging for the information, unlike Prestel, you can browse over the information, saving to disk as you choose. This data option has been one that has been strangely neglected by users, to such an extent that the Teletext cards can be found at bargain prices. If you can connect the Teletext card to a good signal source from an aerial this form of data access is very much more useful than Prestel for a lot of the information that is supplied on both systems, and without the cost and restriction of telephone lines.

Tape backup systems

When your computer contains a large number of files that you have created, the value of this software can be considerable, because it represents a large amount of your time. The value might not be apparent to anyone else, but if your hard drive is stolen, destroyed in a fire or by dropping the computer, or if the hard drive simply expires of old age or (the least likely possibility) it is poorly made, you will have lost your work, perhaps the work of several years.

One answer, of course, is to make backups on floppy disks of your data files at the same time as you save them to the hard drive. This is a good solution for small-scale work, but the number of floppies that you need for a backup of 100 Mbyte or more makes it too cumbersome.

Users of Windows 95 have a backup system on board, and this allows the use of floppies or a tape drive. Files can be selected, and your selection saved, using a filename, so that you only have to specify that you want to save a set of files known as MYWORK, for example, to have the selection made and used. You can select a complete directory (folder) so that any new files you place in that directory will automatically be backed up. When floppies are used, the files are compressed, so that files that amount to more than 4 Mbyte on the hard drive will fit on a single floppy.

If you need to backup the whole system, or if you have a large amount of data, the cheapest method is the use of a tape drive, because the prices of such drives have fallen to a very low level. The best-known of these drives are the Colorado Jumbo drives (a division of Hewlett-Packard), and at the time of writing the older 120/250 Mbyte drive had been discontinued after selling for about £75 for some time. The larger sizes are still in production, and though the cartridge prices are high compared to floppy disk, the capacity is also high, at least 350 Mbyte on a cartridge costing around £12.

The tape drive itself is very easy to install. It fits in a 5¼" bay, so that there is no need for adapters, and it connects in the same way as a floppy drive, using the standard power connector and the floppy drive data connector (the one you would use for a B: drive if you had two floppy drives). You can use the software that accompanies the drive if you are working with DOS or with Windows 3.1, but Windows 95 already contains the software (written by Colorado). Windows 95 recognises drives of the QIC 40, QIC 80 and QIC 3010 types manufactured by Colorado, Conner, IOmega and Wangtek; also the Colorado Trakker types that connect through the parallel port. The more modern Travan, QIC-wide, QIC 3020, and SCSI drives are not recognized – if you use any of these you will have to use the software that comes with the drive.

The software provides for formatting. This is a time-consuming process, taking up to several hours per tape, and if time is important you can spend a little more on the cartridges and opt for ready-formatted types (the QIC 80 format). You can also erase a tape, and such an erase is complete – you cannot opt to erase part of a tape.

Using tape backup is quite a novelty after years of disk use. The tape shuttles to and fro for some time before you make any options about backing up, and when you select your files (or use a named set) a further time is needed to sort out what is to be saved. The tape then runs in bursts ('streaming') and if you are using the Windows 95 software you will see a visual report on progress, with the time noted.

If you are simply backing up data files, you will find that a backup takes less then ten minutes, but a full-system backup can take considerably longer. A full-system backup should be done at longer intervals, and you might like to keep one copy in a bank vault. The more you can spread copies around, the safer your data is, and the old adage about keeping all your eggs in one basket is pertinent.

Another possibility is an interface card that allows you to use a VCR for backup of up to 3 Gbyte on an E180 cassette.

Controllers

Most PC users have no need to interface the machine to anything other than a printer and possibly a modem. The machine is capable of much more, however, and it can be interfaced to all sorts of electronics systems if suitable interfacing devices are fitted. The addition of interfacing hardware, however, is not sufficient, because software must also be present to control the way that the computer accepts or delivers data. Many interface devices come with software, typically to allow the PC to be used as a digital oscilloscope, a spectrum analyser, or a digital voltmeter. If you want to use the PC for tasks such as controlling security systems, monitoring temperatures and controlling heating or air-conditioning, controlling TV and audio systems, or whatever applications you have in mind, then you must write your own software. This can be done using what is called assembly language, close to the natural number-code programming system for the chip. Assembly language produces very compact programs, but is not easy to write and test. An easier method is to use one of the many higher-level programming languages such as QBASIC, Turbo-Pascal, or C. Such programming is beyond the scope of this book, and only a hint of it can be provided here. The basis of digital outputs is that a byte of data held in a register (a memory unit) of the microprocessor, can be copied to a port by a command such as OUT, specifying the port reference number. If for example, you want the lowest-order digit in a set of eight to be at the 1 value (to turn something on) then you would use a command such as OUT 1 (specifying also the port number). Using OUT 2 would set the second digit, and OUT 3 would set both the first and second lines to be switched on.

Using the PC as a controller requires A-D and D-A converters, of which the A-D type are more common. These allow the computer to be used for data acquisition so that analogue signals input to the converter can be used to display or print information. When the computer is used for controlling systems as well as acquiring data, outputs will be needed, but these seldom need to be analogue. For example, if you are using the computer to control a heating system, the analogue inputs from thermostats will need conversion to digital form, but the outputs will be used to switch devices such as circulating pumps, boiler supply, motor-actuated values and so on. These require a digital output for each device with a simple interface which will basically use a relay or Triac with amplification for the small (0 – 5V) digital signals and isolation so that there is no

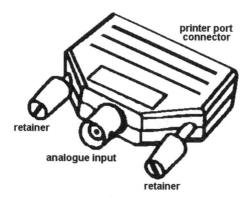

Figure 11.4 *The Pico ADC-10 analogue to digital converter which fits on the parallel port connector of the PC. Other models permit a greater number of digital bits to be used in the conversion.*

possibility of connecting PC circuits to mains voltage – interfaces should use transformers or optical couplers to ensure complete isolation.

One very popular range of interfaces is available from Pico Technology Ltd. and one model in the range (Type ADC-10) is retailed also by Maplin. The ADC-10, Figure 11.4, fits into the parallel port of the computer, so that unless another parallel port is available the printer cannot be used at the same time. The unit can draw its power from the port connector so that no external supply is needed. An analogue input can be connected to the single BNC connector on the unit, and the input voltage will, under software control, be converted to 8-bit digital form. The normal input voltage range is 0 to +5V, with overload protection for ±30V. The rate at which information can be sampled depends on the clock rate of the computer, and is quoted at 10 kHz for an 8088 machine up to 25 kHz for a 33 MHz 80386 machine. The accuracy of conversion is, as is normal for A-D conversion, one least-significant bit (one part in 256 or about 0.4%).

The software that is supplied allows the computer to be used as a single-channel storage oscilloscope with controls for timebase, trigger, and multiplier, and provision for displaying notes and ruler scales. The normal VGA card is supported along with Hercules, CGA and EGA, and printing can be directed to Epson FX and LQ dot-matrix printers or to the H-P Laserjet. The screen output is illustrated in Figure 11.5, with the

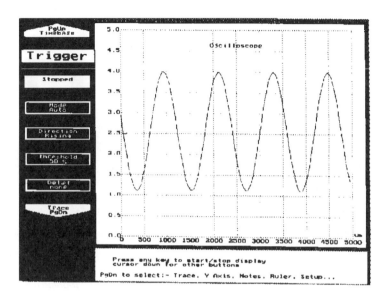

Figure 11.5 *The screen appearance of the oscilloscope display (the reproduction is poor in this copy, but the general layout is as shown).*

grid lines and scaling figures which are helpful in using the display to measure amplitude and timings. The 'buttons' at the left-hand side of the display can be clicked to bring up the action that is noted on the button. The Page Down key on the keyboard will select other options. The timebase can be set from 1 ms per division to 5 s per division, with single, auto or repeat triggering. You can zoom in on any particular part of a trace to see more detail.

The spectrum analyser software allow control of min/max frequency and averaging, with a grid display and optional titling. This is illustrated in Figure 11.6, showing a display of signal amplitude plotted against time. You can set the start and stop frequencies that you want to use and add notes. The traces can be saved and printed (note that if you have only one parallel port you would need to save the trace for subsequent printing).

Voltmeter functions, Figure 11.7, include min/max values, decimal places, and units, and a title can also be added. The digital output is illustrated here, but an analogue bar-graph display can be used – there is no option for a needle analogue display. You can select whatever scale you need, though you need to be careful of overloading the input to the

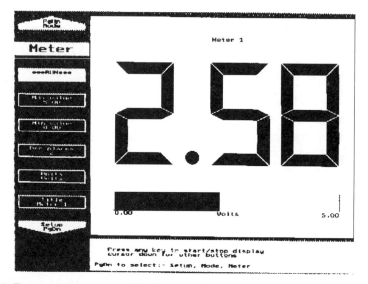

Figure 11.6 *The screen appearance of the digital voltmeter – note the bar-graph display under the digits. The actual screen appearance and the printed version look considerably better than this print indicates.*

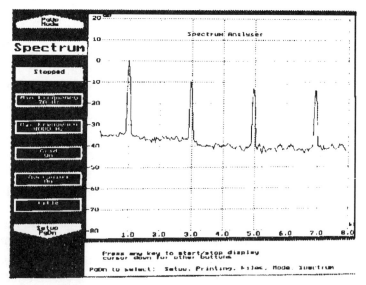

Figure 11.7 *The screen appearance of the spectrum analyser display, again rather poorly reproduced in this copy.*

converter, and an attenuator will be needed if you deal with voltages higher than the usual TTL range of 0 to +5 V.

Currently there are three other products from Pico Technology that deal with data logging. The ADC-11 is a 10-bit 11-channel A-D converter that, like the others, plugs into the parallel port socket. The analogue inputs use a 25-pin D-type connector on the unit, and one digital output is provided so that a single control signal can be obtained. The software consists of the PicoScope set of oscilloscope, voltmeter and spectrum analyser, but the oscilloscope is a two-channel type which can be triggered from either displayed channel or a third channel as required. The voltmeter allows up to 11 simultaneous displays to be monitored. The PicoLog software package is used for data logging, with each recorded sample selected as the minimum, maximum or average of a number of readings that are collected over a period that can be as short as one millisecond or as long as one day. The samples can be scaled, using your choice of linear interpolation, table lookup or a mathematical formula. You can arrange an alarm to sound if a value goes out of limits that you preset. Text and graphical reports can be prepared, edited and displayed during and following the collection of data. Sampling rates up to 15 kHz can be used, depending on the clock rate of the computer. Note that if you write your own software you should use either assembly language or a compiled language (Pascal or Basic, for example), but not a language that uses an interpreter such as QBASIC or GW-BASIC.

The ADC-12 A-D converter also plugs into the parallel port socket and is a 12-bit type single-channel for higher resolution. A collection rate of up to 18,000 samples per second can be used (given a high enough clock rate of your computer), and the resolution is one part in 4096. The analogue input is by way of a single BNC socket, and the PicoScope and PicoLog software, as described above, is supplied as standard.

The ADC-16 is a high resolution 16-bit that gives a resolution of 1 part in 128,000 over a voltage range from -2.5V to +2.5V. Unlike the other ADC units it fits to the *serial* port rather than to the parallel port, freeing the parallel port for a printer. No additional power supply is needed, and since the ADC-16 can be plugged in at the end of a long serial cable you can place the converter unit itself near the equipment that is providing the analogue inputs. Up to 8 single-ended input or 4 differential inputs can be used. The supplied software is the PicoLog allowing a maximum of 65,000 samples per run, with a sampling rate ranging from 1 ms to 1 day. Both input and output connectors are 25-pin D-types.

Digital and D-A interfaces

The R-S catalogue (or Electromail) lists a number of boards that can be used for interfacing if spare slots are available. These are manufactured by Arcom Control Systems Ltd, from whom further information can be obtained, and the following is a summary of a few of these boards. You will need to write your own software for these boards.

The PCIB40 board is a 40-channel digital input/output board that allows digital information to be read and/or written by way of a 50-pin D-type connector. Each channel uses open-collector outputs that can sink up to 24 mA at 5V, and each channel can be set high, low or floating (in sets of eight) when power is switched on. The board provides three 16-bit counter-timers, with one providing interrupt signals for scanning channels.

The PCAD 12/16H is a 16-channel 12-bit A-D converter which can handle analogue signals with a maximum frequency of 40 kHz. Conversion can be controlled by software, by an external input voltage, or in sequence using the on-board programmable rate generator. Inputs can be in the range 0 – 0.5V, 0 – 1V, 0 – 5V or 0 – 10V, and the maximum resolution is 0.122 mV per bit. The gain can be selected in a four-decade range by software (some software examples are supplied) and double-buffering is used so that conversion and reading can be carried out at the same time. The connector is a 50-pin D-type.

The PCDAC 12-4 is a 4-channel 12-bit D-A converter, whose output analogue signals from each channel can be in the ranges 0 – 5V, 0 – 10V, ±5V or ±10V. Each driver can supply 20 mA, and software examples for a variety of applications are supplied.

RS/Electromail can also supply a huge range of signal interface and conditioning boards, including inputs for thermocouples, opto-isolated inputs, relay outputs, analogue insulated inputs and solenoid-driver outputs.

Networks

Networking means connecting computers together so that they can share resources such as printers and hard disks. This allows any user of a machine in the network to print to the one laser printer in the system without such expedients as placing the file on a disk to load into the machine that is connected to the printer. It allows users of a database to update files, knowing that these same files are available in updated form to each other user. It allows a document to be seen on each screen in the network if it is important to have everyone's attention drawn to it. Networking promises, for a number of users, a considerably easier life than the use of separate machines.

The promise is not always easily fulfilled. Successful networking requires a mixture of hardware and software that must be adequate for the purpose, well installed and explained thoroughly to all users. Few experienced users of networks who now feel confident in their use would care to go through the initial stages of networking all over again. Imperfection in networking can mean losing files, programs running slowly, endless error messages on programs that have been previously totally reliable.

The problems are notoriously difficult to locate, and the blame for problems is shifted around faster than a hot potato. Such problems do not necessarily point to a network system being a bad one, because there are so many factors involved in a network as compared to the simplicity of a one-user one-machine set of systems. Network manuals are not famed for being written in clear English or for dealing with the problems that users so often experience. Networking may, in some cases, not even be appropriate, and users could obtain all that they want by passing disks around and by the use of a printer switch.

Fortunately, many of the problems that attended the pioneers of networking have now disappeared. MS-DOS 6.22 and Windows 3.1 are very much more suited to networking than the older versions ever were, and the software writers are more experienced in coping with network use. The extent of the problems can be gauged by thinking about a few simple examples. What happens, for instance, if two network users require the printer at the same time or want to alter the same data file at the same time? Suppose that some files are to be allowed to be used by only a few users and hidden from others? How does one user place a message on the screen of another user or use a Windows program that is located in another computer?

All networking systems require additional software to cope with these tasks of allocation and priority that MS-DOS was never intended to solve. In addition, there is the hardware task of wiring up machines to each other, because there are several method of connecting machines, and a networking card will have to be present (except for some very simple systems like Interlink or The $25 Network) in each machine that is connected. A really extensive network can be very costly and require at least one machine, the server, that is quite exceptionally fast. In this sense, a server is a computer which is directly connected to a resource such as a hard disk (the file-server) or a printer (the printer-server), and it is quite common for this machine to be connected to all the important resources.

A server can be dedicated or non-dedicated. A dedicated server does nothing else other than supply the other machines, a non-dedicated server is used like any other machine in the network with an operator keying in data and looking at the screen. If this operator should absent-mindedly switch off, the whole network goes down. The non-dedicated server will need more software in memory than the dedicated type, so that it may not be able to run large programs, and is likely to run slowly. A dedicated server can make use of all its memory for a network operating system and does not necessarily need to use MS-DOS at all. For a network of more than three users, this is almost always a better approach. Another form of non-dedicated server is the peer-to-peer network in which each computer contains enough networking software to act either as a server or receiver, and no machine is tied up because another one is using its files.

The smallest and simplest networks make use, for connections, of the serial ports on each computer, limiting them to the maximum rate of data transfer of a serial port and requiring no extra cards (unless there are other demands for the serial port). This type of system can be used for printer sharing and for limited file transfer, but is not up to the task of, for example, allowing a database to be shared by several users. For the less-demanding applications, however, the use of the serial ports is an attractive low-cost option and it can make use of low-cost cables. Another option is to use the same software with parallel port connections, which makes the network much faster.

The next stage in complexity is a network in which each computer is fitted with an expansion card whose circuits provide for much faster transfer rates than can be obtained from serial ports (10 to 50 times faster typically). The connection is usually by four-core (twin-twisted) cable using connectors that are now standardized. Where the maximum cable length is less than a few hundred metres it would be pointless to use any more expensive cable system (such as data cable), and the software of the system is usually geared to the less-demanding uses.

The networks that are at the top of the range are all named varieties which are by now well known and well established, such as Ethernet, Token Ring and Novell. For large-scale users, the Novell network system is almost an automatic choice, particularly if the server is to be a mini-rather than a micro-computer. Now that Novell has linked with Digital Research, DR-DOS contains some networking software.

In general, if you are considering networking a large number of computers with any of the major network systems, you will need either to be very experienced yourself or you will need to take advice or have the work carried out by a qualified agent. What follows is advice on installing and using small-scale networks that will serve two or three machines. The systems are simple enough for the user, as distinct from the computing professional, to install and use. The most obvious to start with is Interlink, because it comes as part of MS-DOS 6.0. This is described here in more detail than other software because it is likely to be of interest to a large number of users and is available to all users of MS-DOS 6.0 onwards.

Interlink

Interlink is intended as a simple utility for connecting a laptop to a desktop computer, but it can be used as an elementary form of network system for two desktop computers, using either parallel or serial ports. Most computers possess only one parallel port which is used for the printer, but it is quite common to fit two serial ports and even if one of these is used for a serial mouse this leaves one spare for Interlink connection. If you need faster networking, you can fit additional parallel ports to each computer.

As a simple networking system, Interlink is not quite so versatile as the well-known $25 Network or Little Big LAN, but it is supplied free as part of MS-DOS 6.0 and needs less work to install. The simpler interconnections do not rank as a network for the purposes of installing Windows or MS-DOS.

Serial links, which are much more common for Interlink, allow transfer rates of just over 115,000 bits per second, which is considerably slower than can be achieved with parallel ports, but fast enough for printing or file copying. The hardware consists simply of a seven-core serial non-modem cable terminated in the type of plugs that the machines use – either a 25-pin or a 9-pin type.

Serial cables that are intended for printers or external modems are useless – their connections are not suitable. You should specify that you want a *non-modem* (or *null-modem*) connected cable for linking two computers together. If parallel ports are to be used, the cable that is used to link the computers must be made or supplied specially – it is not a standard form of printer cable (it uses four data pins and four control pins on each connector, plus earth).

With Interlink in use two machines can share drives and printers. One machine is designated as the *server* and the other as the *client*. The client can be used normally, with the drives and printer of the server at its disposal. During this time the server is immobilized – you cannot use it normally. This reflects the main purpose of Interlink, which is to connect a desktop machine to a portable – it is not a peer-to-peer network.

Interlink can be used in a variety of configurations, and you need to decide for yourself, after trying some out, what will be best suited for your own uses. Interlink can be used along with Windows and DOSSHELL on either machine, though task switching cannot be used while the machine is being used as a server.

The minimum requirement to use Interlink is to add INTERLNK.EXE in a CONFIG.SYS line to the machine you will use as a client. You might want to use this line in the CONFIG.SYS files of both machines to allow yourself some flexibility. A typical CONFIG.SYS entry is:

```
device=c:\msdos\interlnk.exe
```

which will install Interlink (whose files are, in this example, in the MSDOS directory) and look for a connection as the computer is starting up. This connection need not exist at that time and can be started at any time later by using INTERLNK or INTERSVR: see later.

The options that can be used following INTERLNK.EXE are:

/**DRIVES:**n   Interlink assumes three drives on the server. If you have fewer or more on the *other* computer you can specify them as /DRIVES:2 or /DRIVES:4 for example. If you use /DRIVES:0 no drives will be shared, only printer connections.

/**NOPRINTER**
              Ensures that the printer is *not* shared. Remember that if a printer is shared both computers must be switched on and running Interlink before printing can be done from the machine which does not have the printer connected.

/**COM**   Can be used in the form /COM:1 or COM:2F8 – the first form is more useful and intelligible. This allows you to specify a serial port (usually in the range of COM:1 to COM:4) rather than have Interlink find it for itself.

/**LPT**   Can be used in the forms /LPT:1 or LPT:378 – the first form is more useful and intelligible. This allows you to specify a parallel port for data transfer (*not* for the printer).

/**AUTO**   Creates a link only if the server is active, otherwise does not load Interlink into memory.

/**NOSCAN**   Installs Interlink in memory, but does not attempt to link with the server. This avoids problems if the server is not switched on.

/**LOW**   Installs Interlink into conventional memory rather than the default upper memory.

/**BAUD**   Used to specify a serial transfer rate if for some reason the normal rate of 115200 cannot be used. The slower rates are 57600, 38400, 19200 and 9600 and they will make access noticeably slower.

/**V**   Prevents timer conflicts – use this if, for example, a serial mouse stops working when Interlink is running.

When the INTERLNK.EXE line runs in the CONFIG.SYS file it establishes a set of new drive letters, equal to the number of drives on the other computer that can be shared. If the number of drives on the other machine is not the default 3, you should use the /DRIVES option to correct this. It is also useful to add the /NOSCAN option if you often start one machine alone or at a different time from the other. If you do not need to share a printer then add the /NOPRINTER option because this reduces the memory requirements of INTERLNK.

The /COM option can also be useful, particularly if you use a serial mouse on the COM1 port. Using /COM:2 in INTERLNK.EXE prevents the program from checking COM1 and possibly interfering with the mouse action.

The normal use of Interlink involves starting the INTERSVR program on the server machine and INTERLNK on the client. There is an option which allows INTERSVR on one machine to copy the INTERLNK files to a machine which is not equipped to load files from a disk, such as a diskless laptop machine which retains files in its memory. Only machines that use MS-DOS compatible files can be used in this way. In this chapter we shall concentrate on the more normal method in which both machines

```
Microsoft Interlnk Server Version 1.00

        This Computer          Other Computer
          (Server)                (Client)
            A:
            B:
            C: (229MB)
            F:
            G:
            H:
            I:
            J:
            K: (130Mb)
            D:
            E:
            LPT1
```

Figure 11.8 *The appearance of the INTERSVR screen when the program has been activated but before any drive is selected. The corresponding drive letters for the other machine will appear when a drive selection is made.*

use the INTERLNK line in the CONFIG.SYS file. This allows either machine to be the server and the other the client.

On the server machine, run INTERSVR, either direct from DOS or by way of DOSSHELL or Windows. This will show only the outline window when the other machine is not yet activated, Figure 11.8.

On the client machine, start INTERLNK either at the MS-DOS command line or by clicking on the command in DOSSHELL or Windows. If you are running DOSSHELL or Windows you can click on a drive letter that is linked to the other machine instead of running INTERLNK. If you have previously used INTERLNK under DOSSHELL there may be remote drive information held in memory, so that clicking on the remote drive letter produces an old directory. Use the F5 key to refresh the display.

The screen of the client machine will then show a list of equivalent drive letters and the display on the server machine will change to reflect the connection. The connection remains in use until it is broken by pressing Alt-F4 at the server .The server diagram remains on screen as a reminder of the connections. In the example, the C:\ drive of the server machine is referred to as F: on the client machine, so that a command such as:

```
COPY F:\words\*.TXT C:\text
```

can be used. For most purposes it is easier to use DOSSHELL or Windows File Manager to carry out such actions. Interlink can be established, once the server is active, by using a remote drive or by using the INTERLNK command at the client machine – when you use DOSSHELL or Windows you need only click on a remote drive letter.

Running INTERLNK from DOSSHELL or Windows presents no problems, but when INTERSVR is run the INTERSVR menu remains on screen. You cannot make use of DOSSHELL actions like task switching while INTERSVR is running, and the same applies to Windows. A message will appear to remind you of this.

INTERSVR offers a similar range of options, adding the /X=B: type of option to specify a drive letter that will *not* be linked. The /LPT, /COM, /BAUD and /V options are as for INTERLNK. The /B option allows the INTERSVR screen to be seen in black and white if you have any problems with the colours. /RCOPY is used to copy the INTERLNK files to a client which cannot load them in any other way.

There are some variations on the Interlink theme. You can force a drive on the client to be equivalent to a drive on the server by using the command INTERLNK clientdrive: = serverdrive: such as:

```
INTERLNK E:=A:
```

You must use the drive letters that are shown when the INTERSVR program is running on the server machine.

You can cancel a link by using INTERLNK clientdrive:= with nothing equated, such as INTERLNK F:=. You may have problems in some cases with different MS-DOS versions. Normally, provided that one machine is using MS-DOS 6.0, the other can be using any version from 3.0 onwards. This is *not* true if the server machine uses a single large disk partition, as is normal, and the client is using MS-DOS 3.0 to 3.4, because these versions cannot make use of large disk partitions. In general it is always more satisfactory if both machines are running modern MS-DOS versions.

If you use the client machine to run a program that is located on the disk of the server you need to be sure that the program is one that could run on the client (not configured for a different video system, for example). In addition, you cannot use the commands:

| CHKDSK | DEFRAG | DISKCOMP | FDISK |
| FORMAT | SYS | UNDELETE | UNFORMAT |

Drives created by a hardware network are not handled by Interlink.

Program types and problems

Programs for the PC type of machine encompass a wider range than for any other computer, because of the long time for which the PC type of machine has been available and its compatibility. In this section we shall look very briefly at some main types of programs and what they do, but not in detail because a detailed description of just one major program, such as Microsoft Word, would require a book larger than this one. Appendix D contains a list of books that deal in detail with some of the latest software packages.

Before we look at the main program types, though, a word about installing programs might be useful. At one time, programs were simple and the PC machine was equally simple. Installing a program was seldom necessary because programs were either run from a floppy disk, or the files copied from the floppy to a hard drive. Typing the name of the program would then run it.

Nowadays, things are more complicated. Programs are specifically intended to have their files held in the hard drive, and the files that come to you on floppy disks are usually compressed and unusable by themselves. In addition, the program needs to be tailored to the machine, usually by typing answers to questions on a form that appears on the screen – these answers are then held in a file (an initialization file) that the program reads and uses each time you start it running.

All of this is too much to expect of a user, so that such programs have a SETUP or INSTALL program which carries out all the actions automatically – you simply have to type some answers as noted above. Though the action of such programs is by now becoming reasonably standardized, you should read with considerable care the part of the manual (often a separate part) that deals with installation. Very often, changes to the machine (such as changing the graphics board) require you to run the SETUP program again.

From then on, you have to learn how to use the program, and in this respect, books are often more useful than the manual. The manual is intended to be a reference guide, not as a primer for the newcomer, and you can only too easily become tied up with irrelevant detail if you try to learn from the manual. In this respect, programs that run under Windows are at an advantage because there is so much that is common to all programs.

Typical software packages

Word processing is the task for which the majority of computers are used for the majority of their lives, simply because some 80% of the activity of a business is concerned with text in the form of letters, reports, memos and the like. The principle is that text is typed so as to appear on the monitor screen rather than initially on paper. In this way, mistakes can be corrected before being committed to paper, representing a huge saving of time as compared to the old practice of typing a report over and over again until it was perfect. No document should ever need to be typed from scratch again, because a perfect copy can be held and used as a template. The use of the screen also makes it possible to wrap words, so that no word is ever split between lines nor is any line longer than a set amount. Figure 11.9 shows the screen appearance of Word for Windows.

The typing and editing of text relies on the use of a cursor, the flashing block or bar whose position on the screen indicates where the next typed character will be placed. The cursor can also be moved independently of the typing action so as to locate any character in the text, using the cursor

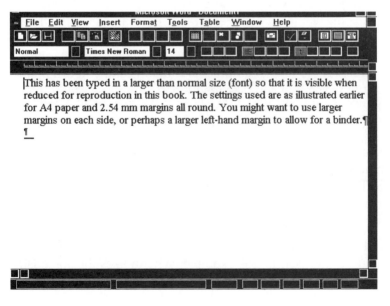

Figure 11.9 *Typical screen appearance of Word for Windows, a leading word processor for the Windows system.*

keys. Apart from the typing and correction of text, the basic facilities that word-processing programs offer are saving and loading, editing any part of a document, use of printing effects, block moves, search and replace and page layout control. Text can be saved on the disk, sometimes automatically after a specified number of characters have been typed, so that anything that you have typed need never be typed again as long as you can locate the correct disk file again. Text from different files can be merged so that a report, for example, can be put together (or boilerplated) from a set of smaller reports and memos that exist on the disk.

Desktop Publishing, or *DTP*, is the name for low-cost publishing, using a computer and its associated printer to prepare material which can be used as a master for further printing work, or to make a limited number of copies of the material directly. The use of a computer allows typography to be handled in very much the same way as word processing is handled, and this is reflected in some word processors (like Word for Windows) in which typographical techniques like the use of different fonts are used. The form of the printed material can be seen on the screen and manipulated as much as you like in this form without a single mark being made on paper.

Each page of the work can be completed and recorded on disk, and only when the whole set of pages is ready need anything be printed. The page can contain text that uses different forms of type (different fonts), in different sizes that allow you to have headlines, sub-headings, main text and notes, along with graphics illustrations that you can prepare for yourself or which you can take ready-made from a selection that comes with the Desktop Publishing package or on additional disks. The pictures from these sources can be placed into the page, with the words of the text making way for them and arranged around them as you choose. Figure 11.10 shows the screen appearance of Page Plus, one of several low-cost DTP packages for the Windows system.

A *spreadsheet* is a form of electronic grid display, Figure 11.11. The display is divided (without any dividing lines necessarily appearing on the screen) into cells (also called slots or fields), and each cell can be referred to by its letter and number co-ordinates, some of which have been placed into cells in this example. What makes the spreadsheet so important and useful is that the contents of one cell can influence the contents of any other (as programmed by the user). This allows a change in the contents of one cell to cause changes in many other cells, so that the whole appearance of a worksheet can be changed by altering one

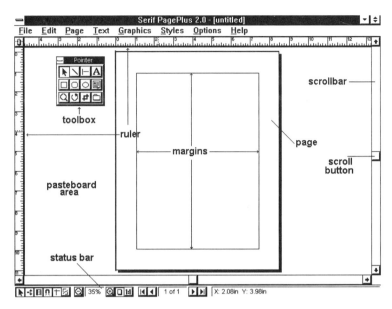

Figure 11.10 *The screen appearance of the DTP program, Page Plus 2.0.*

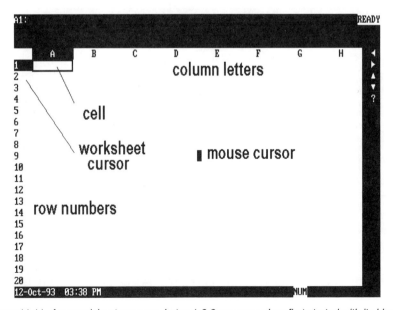

Figure 11.11 *A spreadsheet program, Lotus 1-2-3, as seen when first started with its blank worksheet.*

entry. Each cell, slot or field can contain two different types of entry. The entry that is visible when you look at the worksheet is a number or a piece of text, and this might be typed into place as you enter data, or it might arise as the result of some earlier action. The other, invisible, content of a cell, slot or field can be a formula of some type which will decide what the visible entry shall be. This formula entry can be displayed, edited and erased like any other entry, but is never visible during the normal use of the worksheet.

An example will make this clearer. Figure 11.12 shows part of a worksheet in which the second column consists of the raw price of an item and the third column contains percentage local tax rates. The fourth column shows the result of multiplying the number in the second column by the percentage in the third column. Now when data is entered into this portion of worksheet, only the items in the first three columns are entered: the names, numbers and percentages. The fourth column requires no entry, because the figures that it shows have been calculated automatically from the figures in the second and third columns. This has been done by placing into the cells of the fourth column the formula for multiplying the second column value by the third column percentage.

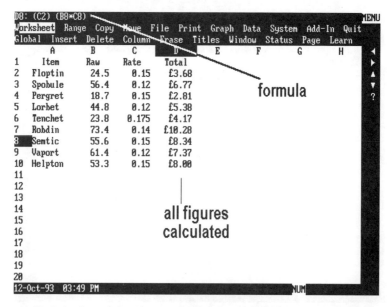

Figure 11.12 *A simple worksheet in use, showing the calculated column and the form of formula that each cell in column D contains.*

The fourth column, though it contains this formula, does not display the formula, only the result, with a zero displayed if there are no entries to work with. The fourth column has also been formatted so as to show the pound sign and with only two figures following the decimal point. Any change in the figures in either second or third column will cause a change in the fourth-column figure, so that the worksheet figures form a dynamic display, always remaining updated and in correct relationship even when alterations are being made.

Most spreadsheets are arranged so that this recalculation action is automatic and takes place each time an entry is made or altered. Note that it is not essential to have a formula in each cell, or in any cell, only where some form of automatic entry is required.

A *database* program is another fundamental business program that is concerned with the filing and retrieval of data. You can design a screen form that allows data to be typed in, with restrictions on space if necessary, Figure 11.13. Once data has been entered, for example, it's very easy to arrange it in the way you want. You can print out the contents of the database, for example, in alphabetical order of words that are used, Figure 11.14. You might, for example, use the database to keep stock

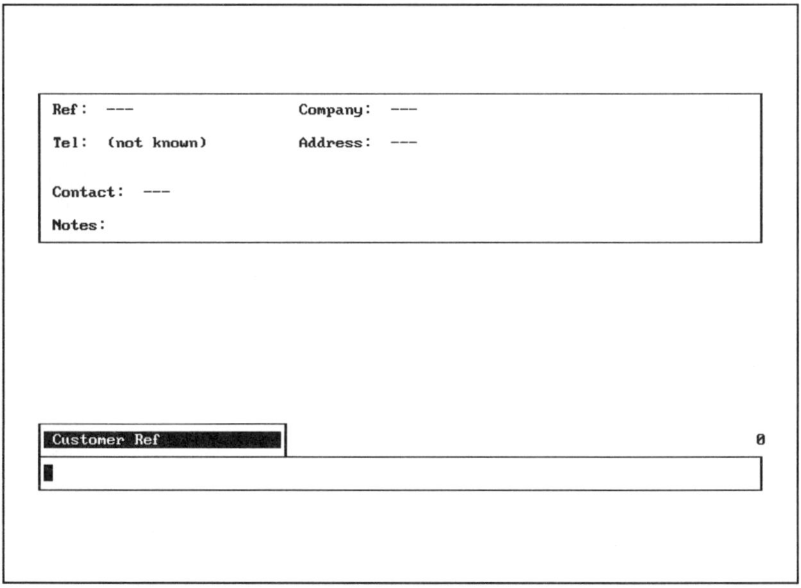

Figure 11.13 *A form designed for a database (Masterfile PC) – you have full control over the design of this form.*

```
Ref: BE              Company: Brownhills Engineering

Tel: 0433 433555/6   Address: Unit 4
                              Claverley Ind. Estate
                              Redstone
Contact: Harry Hughes         Staffs WV3 5HD

Notes:
```

```
Ref: FHW             Company: Freddie Hall-Winter

Tel: 0332 871123     Address: Winter House
                              High Road
                              Standwell
Contact: Selina               Derbys DE3 0YR

Notes:
```

```
Customer Ref           Record no. 1       Length : 121    Sets:
                       [Space]=menu
Data records : 5       Selected : 5       Format 2: Index card style
                                                    General help = F1
```

Figure 11.14 *Displaying data from a file that uses the form illustrated in Figure 11.10. You can determine what records you display by specifying criteria for selection (place, date, alphabetical position etc.)*

records, and you can then print out in alphabetical order of stock item names. You might, on the other hand, like to print out in increasing or decreasing order of stock value for each item, or in order of stock level so that the list became a re-ordering reminder.

The other aspect of a computer database is searching and listing. Any item, a name or a number, can be searched for and a list made of each place where that item is found. The criteria for the list can be specified, so that if you want to search for all the left-handed red-haired electronics engineers in East Anglia you can do so if your database contains such information (which would certainly require registration under the Data Protection Act). As is usual for business computer programs, once you get to grips with a program, you tend to find more and more uses for it.

Graphics programs for the PC are of three types, the business graphics type, of which Lotus Freelance Plus is a prominent member; the draw or paint programs that are aimed at producing artistic work, such as Windows Paint or PC Paintbrush; and the computer aided design packages for technical drawing work such as AutoCAD and AutoSketch.

Programs that aim at producing artwork for business use are concerned mainly with the drawing of graphs from business statistics. These programs are quite often not independent; they are often incorporated into spreadsheets, for example. Data can be linked to charts so that when the data changes, the charts will be altered also. Word-processing abilities allow the preparation of text charts, allowing for bullet marking of paragraphs and indenting. Drawing capabilities include organization charts, flow charts, technical illustrations and diagrams, using shapes such as lines, rectangles, circles and arcs. Closed shapes can be filled with colour or hatching, typescript can be added, and all of the graphics elements can be shrunk or enlarged or stretched in one direction only.

Painting programs can make use of images that you create for yourself, or images that have been taken from other sources. Windows Paintbrush provides an excellent illustration of the principles. Lines, boxes and circles can be drawn, using filling for closed shapes if required. Any piece of a drawing can be moved, copied or saved to be used elsewhere, and text can be added as required. Any part of a drawing can be edited in terms of the dots that make up the screen picture.

The term CAD means Computer Aided Design (or Drawing). The work of the draftsperson is to create drawings which will later be used in the manufacture or planning of anything from industrial components to kitchen layouts and add-on conservatories. The essential difference between a drafted plan and a drawing is that a drafted plan is to scale. If, for example, you are planning the layout of a kitchen, knowing that the floor space is 4.5 m x 3.5 m, you need to have the shape of the kitchen drawn out to scale, perhaps a scale of 25:1, so that 4 cm on your plan represents 1 m in your kitchen. The kitchen appliances will then be drawn to this same scale so that their position in the kitchen can be accurately represented on the plan. With such a system, you could then make cut-out drawings of the appliances and move them around the plan to determine the best fit. Possibly not all of the appliances could be moved in reality, so the main plan might show the fixed appliances such as the boiler and the sink, leaving you free to pick the positioning of others.

The point of using a computer to aid design is that much of the tedious work can be eliminated. Setting a scale, for example, becomes no more than typing a figure. Placing lines on a plan is done with an image on the screen, so that the plan does not have to be printed until the lines are perfectly positioned. The system can be arranged so that lines can be drawn approximately and will snap into their correct dimensions and

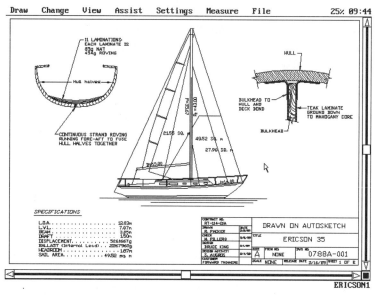

Figure 11.15 *The screen appearance of the Autosketch CAD program, showing one of the examples.*

places. Text can be added to the plan so as to display dimensions, scales, label points and so on. Taking the kitchen plan example again, the kitchen can be drawn, and the appliances drawn separately (on another layer) and moved around with no need for paper cut-outs, until they fit the way that you want. You can then print the plan and think about it, perhaps to make use of it or to come back to the computer and modify it. Most important of all, you can save the complete drawing as a disk file and return to it again and again, making other copies, editing it, re-printing as you wish. CAD bestows the freedom for designers that a word processor gives to authors. Figure 11.15 shows one of the AutoSketch examples.

Section **V**

Speed and power

Chapter **12**

486, Pentium and beyond

What if you need to keep up with the leaders of the pack? If you want to use modern software, which implies Windows 95 and the software that is designed to be used with it, then a simple machine such as has been the subject of Sections I and II is not adequate. Here's what is needed:

- A processor in the 486 or Pentium class. A 486 needs to be operated at 66 MHz or more, preferably 100 MHz. A Pentium should be used at 75 MHz or more. The prices of Pentium chips and motherboards are falling rapidly at the time of writing.

- At least 8 Mbyte of fast memory, and preferably 12 Mbyte. In addition, this memory must be of the 72-pin SIMM type, not the older 30-pin type.

- At least 512 Mbyte of hard drive. Machines of this standard are being supplied with 850 Mbyte drives as standard, with low-cost upgrades to larger drives.

- A local-bus system of the PCI type, rather than VLB. There's nothing wrong with VLB, but all the most recent Pentium motherboards use PCI.

- The EIDE type of drive interface, allowing modern hard drives and CD-ROM drives to be used.

- A fast graphics card, optimized for Windows 95, with at least 1 Mbyte of RAM.

- A monitor that is capable of 1024 x 768 resolution – this does not mean that you are obliged to use a screen of this resolution, but the capability should be there.

If all this looks rather depressing, remember that this is the price of progress. Today's top-class machine is tomorrow's entry-level, barely-adequate one and if you need to keep up you have little choice. Since the early part of this book has illustrated how little is needed for essentials, this section will concentrate on performance. At the time of writing, suppliers in the High Streets of towns could supply such machines for £850 or less, though even at the time of writing many were still trying to sell machines with only 4 Mbyte of memory as 'suitable for Windows 95' or 'with Windows 95 packaged'. By this time, no one should be trying to sell a machine with less than 8 Mbyte of RAM as suited for Windows 95.

Upgrade or buy?

As in the earlier parts of this book, you always have the problem of whether you can upgrade to modern standards or not. One point needs to be made right away. It is perfectly true that you can upgrade a processor to 486, simply by plugging in a replacement chip (or, in some cases, an additional chip). All this does, however, is to improve the processor performance. Now this might be all you need, if all the upgrading you want is faster performance on older software, but it is not a route to the use of Windows 95 and the multimedia software of 1996 and beyond. The reason is that an upgrade to the processor is only a fraction of the requirements; it does not deal with a slow video board, slow hard drive, lack of a local bus, or lack of memory. These factors are equally important if you want the upgrade to be really useful. The only upgrade of this plug-in type that is really useful is when you have a 486 machine with a Pentium socket and this is justified only if you can obtain the correct version of the Pentium (often a P24T type) and if the rest of the machine can match the performance of the upgraded processor. Remember that you may need to contact the maker of the motherboard to find out what Pentium needs to be used (from what appears to be 57 varieties). In addition, the cost of the upgrade can be considerable until chip prices fall (which will be about the time that you will want to change to something faster still).

If you are presently using a 386 machine, such as is the subject to Sections I and II, there is no easy upgrade to 486/100 or Pentium performance other than the fitting of a new motherboard. At the time of writing, Pentium motherboards were ridiculously overpriced, often only £200 less than the price of a complete machine (including monitor) of equivalent specification. This situation will not be permanent, but no one ever has time to wait, so if the Pentium motherboard that you want would make the whole upgrade too expensive, consider buying a complete machine from a source such as Escom (if you want to see before you buy) or any of the advertisers in magazines such as *PC Plus*. If you buy from Escom, make sure that the package includes a mouse.

Upgrading is, in any case, often dubious. How large a hard drive does your older machine use? Few 386 machines used large hard drives, and virtually none of them used a local bus along with an EIDE hard drive interface. Even if you have a 386 with a SCSI type of interface and a large hard drive, how much life remains? A large hard drive on a 386 machine is not likely to be a *young* hard drive, and it will need replacement sooner rather than later. What type of memory is used? A 386 machine will almost certainly use the older type of 30-pin SIMMs, and the new motherboard will use the later 72-pin type. Though you can buy adapter cards that allow you to use a set of 30-pin SIMMs in a 72-pin holder, this is not a complete solution because you will need more memory than the old machine ever used. In addition, a new Pentium motherboard will use a PCI local bus, and your hard drive card and graphics card will not fit this bus; they will have to be used on the ordinary bus with a sacrifice in working speed.

All in all, it makes sense to leap to the new standards by buying a new machine, so what should you look for? You need to ensure that the new machine is itself adequate. This is highly likely if it is a Pentium machine, but there are large number of 486/66 (and lower) machines being sold off at the time of writing and probably for some time to come. These often use 4 Mbyte of RAM or less, and are not always up to date in terms of local bus, hard drive, and other features. The only factors that might make one of these machines a reasonable buy would be a very low price (less than £300) and easy upgrading. Since it's not easy to judge how difficult it might be to upgrade such a machine they are better left alone.

If you don't need the utmost in speed, a 486 type of machine with a VLB local bus, fast graphics card and 4 Mbyte of memory that can be upgraded using 72-pin SIMMs can be a good buy if the price is right. Watch in particular for machines which use the IBM Blue Lightning (BL) chip. There is nothing wrong with them, but they cannot be upgraded quite so easily, and the BL chip is not equivalent to a 486DX, because it lacks a maths co-processor. This is not important if you use only word-processing applications, but it can make the machine slow for CAD or spreadsheet uses. In the past, some advertisements for BL D2 machines have been worded in such a way that you might believe that you were buying a 486DX/66. At the prices asked by dealers such as Morgans, the BL machine is a good buy if you want a fast and capable machine, and its tower casing makes upgrading to multimedia easy, though the Pentium upgrade is more dubious (the socket is not a ZIFF type on the machine I use).

Even if you buy a Pentium machine, you cannot be sure that you are buying into the most recent version. At the time of writing, there was a dazzling variety of Pentium chips and motherboards, and the range of choice is not likely to get less. Two items need to be sorted out, chipset and motherboard.

The *chipset* means the chips that are used along with the Pentium, and which contribute significantly to the performance of the board. At the time of writing, you could buy boards that featured chipsets called *Triton* and *Neptune*. Of these, the Neptune set is the older, and if you buy a machine with this chipset you might need to upgrade sooner. The Triton chipset is the more modern, and you should check that any machine which is on offer uses this set. It's possible, of course, that by the time you read this the Triton set will be old stuff and something else will be top dog (perhaps a Poseidon chipset, if they keep to this type of name). All you can do is to read the magazines to find what's new and what's old. Remember that the price for the most recent equipment is always pitched to allow for the fact that prices fall faster than autumn leaves once something gets into volume production.

In addition to the chipset, you have to consider the design of the board. The reason for this might be temporary, because it arises from a shortage of fast cache memory. All Pentium motherboards need fast cache memory to serve the processor, and the traditional way of doing this is to use static RAM chips (SRAM). These have always been expensive, and at the time of writing they are scarce as well. They are fitted into the top Intel motherboard, known as *Aladdin*. This also uses a new variety of RAM chip called EDO (Extended Data Out). Needless to say, EDO costs (slightly) more than the older type of 72-pin SIMM, but you can mix and match, so that you need have only one EDO SIMM and you can use the older type for the rest of the memory. There is a more recent version of EDO around, called Burst EDO, and it delivers still faster performance.

Because these Aladdin boards are expensive and as scarce as the cache chips they use, there are two other grades of boards. The next grade down is called *Endeavour*, which uses the EDO RAM but no SRAM cache. Instead, the cache makes use of the EDO RAM, and though the performance is not as good as that of the Aladdin board, it is not exactly slow. In addition, the Endeavour board can be upgraded to Aladdin standards by plugging in SRAM cache chips as and when they become available.

Warning: Always take upgrade promises with some scepticism. These promises may be made with good intentions, but they often go awry. There are, for example, thousands of computer users with 486 boards who thought they could easily upgrade to a Pentium using a socket provided on their motherboard, but found that no Pentium chip fitted the socket. The promise was made in good faith, and the makers of the motherboard sincerely believed that a suitable Pentium would be made, but it was not.

The lowest grade of Intel board is called *Zapper*, which has neither fast cache memory nor EDO. It is faster than older boards using the same grade of Pentium, but not so fast as its companion boards.

What if you find a motherboard that carries none of these names? For example, there are still a lot of motherboards around that use the older design called *Plato*. You will simply have to go by the description of performance. Look for SRAM cache on board, and a chipset that is Triton or an equivalent, plus a socket that can be used for upgrading. Ask to see the chip that can be used for the upgrade – if you can't get one yet, make your excuses and leave. If you are shelling out between £350 and £700 for a motherboard alone you are entitled to get the best there is. Remember, however, that what costs £700 now may very well cost £350 in six months time.

Sockets

Any socket that is intended to be used for upgrading a processor must be of the ZIFF type. The letters mean *zero insertion force* socket, and they are easy to recognize because of a lever at the side. The lever opens and closes the connections, so that you can open up and drop a chip in with no need, as the name suggests, to apply any force. You then swing the lever over and the chip is tightly clamped and connected. The older non-ZIFF type needs very great care to avoid bending any of the pins on the chip as you try to ease the chip fully into the socket. Remember that some chips use a lot of pins – if you don't think there is much to getting 168 (or more) pins into their holes without undue force you probably haven't tried it.

Hard drives

The hard drive is a vital part of any computer, and its role has become more important recently. At one time, the use of MS-DOS limited the size of hard drive that you could use to 512 Mbyte. If you wanted to use a larger hard drive you needed to partition the drive into two or more drive letters. For example, if you used a 1 Gbyte drive (1024 Mbyte), then it had to be partitioned as two 512 Mbyte drives, using letters C: and D:, or as three drives, such as 512 Mbyte C:, 256 Mbyte D: and 256 Mbyte E:, and so on. The partitioning was carried out using the FDISK utility that is part of MS-DOS, and since partitioning scrambles all the data on a disk, it had to be done before the drive was formatted, or after a full backup was made, so that the new disk letters could each be formatted.

Partitioning is no longer needed for any drive of less than 2 Gbyte, provided you are using a suitable modern machine with an up-to-date BIOS chip and also running Windows 95. This is why you now find the 850 Mbyte drive as a standard fitting of Pentium machines, and at the same time, the prices of hard drives have fallen dramatically – at the time of writing you now get twice as much capacity for your money as you did only a year ago. There are still a few machines (famous names, too) which use a BIOS that does not allow for large hard drives, and you can either use software (such as *Disk Manager*) that will make up for the deficiency or plug in an EIDE driver card that contains its own ROM with provision for up to 2 Gbyte hard drives.

The need for large hard drives has arisen out of the size of software. Even before Windows 95 came along, programs like Microsoft Office were packaged on 31 floppies, and demanded a lot of hard drive space. Windows 95 typically uses some 85 Mbyte of hard drive and you can expect to need a comparable amount of space for the applications, like Word 7, that you use along with Windows 95. In addition, Windows sets up a permanent swapfile on the hard drive to use when there is not enough memory to allow programs to be swapped around.

Hard drives have changed in the last few years. At one time, the software steps of adding a new hard drive (or a second hard drive) followed the route illustrated in Chapter 3. More modern machines use a BIOS which allows the hard drive to be recognized automatically, so that plugging in the drive automatically installs it, with no need to alter jumpers or to use the CMOS RAM entries. This has two important consequences:

1   You cannot take an old IDE hard drive from a 386 machine and plug it into your new 486 or Pentium machine.

2   You cannot upgrade an old 386 machine with a new hard drive of more than 512 Mbyte and expect it to work, particularly if it is a second drive. You may be able to use it as the only hard drive, but don't bank on it.

The first of these is the more restrictive. Modern machines use the auto-recognition system that depends on a ROM inside the hard drive along with the BIOS ROM in the computer, and they have no provision for setting up disk parameters in the CMOS RAM (see Chapter 6). The older hard drives have no information in ROM, so that when they are fitted, they simply do not register when the BIOS tries to find their type. Since you cannot enter disk data into CMOS RAM, the older hard drives cannot be used. This is another argument for caution when you are deciding to upgrade   buying a new machine ensures that the motherboard and the hard drive match each other.

In addition, you should ensure that the hard drive interface is of the EIDE (Extended IDE) type (unless you use SCSI). This interface is placed on the motherboard in some designs, and if it is not it should plug into the local bus, which, as noted earlier, should be the PCI type. The use of EIDE ensures that you can run up to four hard drives or equivalent devices such as CD-ROM drives.

Using DriveSpace

Since Microsoft introduced DriveSpace (and its predecessor, DoubleSpace) this has been a useful way of getting the equivalent of more space on a hard drive, and it solves the problem of the hard drive which is uncomfortably full. Adding a second hard drive is, even at today's prices, not a cheap option, and it invariable entails a large amount of work in shifting files from one drive to the other. Converting to DriveSpace takes only time, because the system is free with MS-DOS and with Windows 95. An updated Version 3.0 of DriveSpace is one of the items on the Microsoft Plus package that has been released along with Windows 95 (this is, however, a separate item).

DriveSpace is a piece of software that compresses files, reducing redundancy, so that they take up less space on the hard drive (or on a floppy disk). The files are compressed when you save them and expanded again when you load them, so as far as you are concerned, the only difference is that you can get many more files on your hard drive. The compression averages out at about twice, so that your 240 Mbyte drive seems to store up to 480 Mbyte of data. The snag is that the equivalent maximum capacity of hard drive is 512 Mbyte, so that if you compress a hard drive of more than 256 Mbyte you will have to partition it. This is definitely not something that you would want to do with a drive that is filled with data.

If you are using Windows 95, however, you can buy the add-on Microsoft Plus program set, which includes DriveSpace 3.0. This raises the limiting size to 2 Gbyte, so that the usual 850 Mbyte hard drive can be compressed to take some 1.7 Gbyte of files. It's likely that you will already have data on a drive of this size, so that using DriveSpace 3.0 is a logical solution to the problems of limited drive space. If you are using an older small drive, of perhaps 250 Mbyte, the use of DriveSpace 3.0 is almost essential if you want to use Windows 95 without upgrading the hard drive first.

Memory upgrades

Next to upgrading hard drive space, which is often necessary simply to find space for Windows 95 and its associated programs, an upgrade of memory is essential for use with modern programs. All motherboards from 386 onwards use memory in SIMM form, and the type of holder that allows you to insert the SIMM and then turn it to lock it into place is almost universal. What marks out the older boards from the more recent is the way that *banked memory* is organized.

The memory that is used for RAM is dynamic memory, which unlike the static type has a very limited retention time. Static RAM will hold its data for as long as power is applied, but dynamic RAM will lose data in a few thousandths of a second. Because of this, all dynamic memory needs to have refreshing pulses applied at intervals of not more than one thousandth of a second (1 millisecond), and this, along with the time that is needed to read or write the memory, limits how fast a processor can make use of memory.

The normal method of dealing with this is the use of banking and fast cache. Banking means using memory in groups, so that when the processor reads or writes it works with the groups in sequence, not using any one group for more than one byte of data. This gives time for the unused groups to be refreshed and to recover from a read or write action. In addition, if RAM is read or written to or from a fast SRAM cache, the transfer can be done 'in spare time', because the processor is doing its reading from and writing to the cache rather than directly from or to the RAM.

The consequence for the older 386 machines was that you could not plug in memory just as you wished. These older motherboards took 30-pin SIMM units and provided for up to four. The banking was usually in pairs, however, so that the minimum you could fit was 2 x 1 Mbyte for a total of 2 Mbyte of RAM. If you filled the board with 1 Mbyte SIMMs you had a maximum of 4 Mbyte, just adequate to run Windows 3.1. If you wanted to upgrade, you needed to upgrade *at least* in pairs. This usually involved altering jumpers or switch settings, and some machines required all four sockets to use identical SIMM sizes. If, for example, you wanted to upgrade from 4 Mbyte (4 x 1 Mbyte) you had to go to 16 Mbyte (4 x 4 Mbyte) in one step, and you needed to set jumpers or switches to notify the arrangement of memory units. Furthermore, your 1 Mbyte SIMMs were redundant until recently – you can now buy a SIMM carrier which will allow you to stack a set of 30-pin SIMMs on to an adapter which also has 30 pins, so that you can plug in, for example, 4 x 1 Mbyte SIMM units into one motherboard space.

The modern 72-pin SIMM units incorporate banking in each unit, so that each SIMM can be independent of the others. You can, for example, start with one or two 4 Mbyte SIMMs and upgrade with more 4 Mbyte SIMMs or with 8 or 16 Mbyte units as you please. This allows much more flexible upgrading, and you can now buy adapters that allow the older type of SIMMs to be used in sets and plugged into a 72-pin slot. A few 486 motherboards have sockets for both 30-pin and 72-pin SIMMs.

The most recent memory units are EDO (Extended Data Output) types, used on some Pentium motherboards. The EDO memory can be run at faster speeds, and it can be recognized automatically by the operating system, so that you can mix EDO with the ordinary 72-pin SIMM memory.

Graphics cards

Using Windows, either 3.1 or 95, involves using a graphics display at all times, unlike the old days when applications such as word processors could use the fast-working text screen display. This has nothing to do with the monitor; it's dictated by the interface card that sends out signals to the monitor. Many computers come with a graphics card that is politely described as 'adequate', meaning that it takes its time about managing Windows displays, and since the speed at which a display can be changed is often an important factor, this also can be a bottleneck.

The first important point is to use a graphics card that sits in a local bus socket, whether VLB or PCI. If the graphics card is one that uses an ordinary expansion slot, it will be slow no matter what is going on inside the card – after all, a Ferrari isn't cost-effective on farm tracks. As far as the difference between VLB and PCI is concerned, most 486 mother-boards are fitted with either PCI or VLB, but Pentium motherboards are by now all using PCI, so that's the way to go. Graphics cards can be obtained for either VLB or PCI, and the prices are almost identical, but the two standards are not interchangeable.

Your choice of graphics card then boils down to how much you are prepared to pay for the additional speed and memory capacity. Memory is important only if you intend to use the more demanding resolution options. If you intend to use the VGA standard of 640 x 480 with 16 colours (and why not, if you are not using CAD or drawing packages?) then 1 Mbyte memory is more than enough, and you could get by with 512 Kbyte. If you insist on using 800 x 600 or 1024 x 678 resolution, and you might fancy the more demanding colour options from 256 colours to 16 million colours, you need more memory on the graphics card, and 2 Mbyte is the going rate at present.

The top of the range graphics cards use 64-bit processing, and a typical example is the Orchid Fahrenheit Pro64, which is available in VLB or PCI versions, with 2 Mbyte or 4 Mbyte of RAM. The 4 Mbyte version is currently quoted at around £350 (once the price of a complete 386 system). Further down the scale, the Diamond Stealth 64 costs about £90 for the 1 Mbyte version and gives a performance which was a very short time ago regarded as among the fastest.

All of which should suggest to you that you are always firing at a moving target in this business, because whatever is top-notch today is likely to be superseded tomorrow. That's no reason for delay, however, because you always have to balance what is desirable against what is practical. If you upgrade at regular intervals, perhaps every two years, you do not need to worry quite so much about having the ultimate at any given time.

Monitors

Your monitor is not likely to present a brake on the system in the way that a small hard drive or an under-performing graphics card can, but its also important because it's the interface between the computer and your eyes. As displays become more graphical, and particularly if you are tempted to use higher resolution displayed like 800 x 600 or 1024 x 768, the size of a monitor becomes more important. Monitors that are supplied with computers are almost all described as 14" or 15", and it is not easy to tell one size from the other, because it depends where you measure it. The accepted system for TV screens is to measure the diagonal across the tube, disregarding the mask that is placed over a TV tube to frame the picture. Monitor measurements also use a diagonal, but some 15" screens seem to be measured to points that are not normally visible, so that it's not necessarily an advantage to pay any more for a 15" type.

If you use high resolution, DTP or CAD software, or you want to work with a lot of tiled windows, the standard screen size is a limitation and a strain on the eyes. Unfortunately, larger screens are expensive. A monitor cathode-ray tube is made to a much higher standard than the tube for a TV receiver, and because the larger sizes are made in comparatively small numbers they are correspondingly expensive. For example if you pay £150 for a 14" monitor, the price for a 17" monitor of the same make is likely to be closer to £500, and for 20" displays the prices are well above £1,000. The difference a 17" screen makes, however, has to be seen to be believed, and since you do not need to change it when you change the computer it can be a long-term investment. Unlike early models, a cathode-ray tube has a long life (I have a 27" Philips TV tube which is still going well after seventeen years) and it will see out several computers over the years. The more intensive use of Windows is now making the 17" size of monitor more popular, so that prices are starting to decline and a few models are now retailing at under £400. This trend is likely to continue, but the larger sizes are likely to remain expensive because they are mostly sold to buyers (such as newspaper offices) which must have them and who can afford them.

One point to remember is that the larger screen sizes come in larger boxes, so that you have to be certain you have enough depth on a desk, or that you can let the back of the monitor overhang the desk. 17" TV tubes are short, but this is achieved at the expense of picture quality, and such compromises are ruled out for a monitor design you don't have to read 80-column text on your TV screen.

ADI now offers a 17" monitor which can be swivelled through 90° so that you can use it for DTP or word-processing applications that require A4 pages.

Glossary of terms

This is a small glossary that applies particularly to terms used in Windows and MS-DOS 6.0/6.2. For a full explanation of terms used in computing, see *Collins Dictionary of Personal Computing* by Ian Sinclair.

Active icon
The Windows icon which has been clicked on and whose menu will appear on the next click.

Active printer
The printer which will print out from your Windows work. Only one printer is active at a time, though several printers can be installed.

Active window
The window in which you can make entries and select items. Other windows can display on the screen but do not respond to the use of keys until you switch to one of them. Programs can, however, continue to run inside an inactive window, carrying out actions such as searching and sorting which do not require your attention.

Application
A program or suite of programs for a particular purpose such as a spreadsheet, word processor, desktop publisher, CAD program, etc.

Application icon
A Windows icon representing a program that appears outside a menu box on the main screen display, normally at the foot of the screen though it can be moved elsewhere. Clicking on these items will select a program to run, double-clicking will start the program running.

Associate
To nominate a filename extension as one created by an application, so that TXT might be associated with a word processor, SKD with a CAD program, PUB with a DTP program and so on.

Attribute
One of a set of markers in a file which can make the file read-only (the R attribute), Archive (changed but not copied), System (essential to operation of computer) or Hidden (not appearing in a directory listing).

AUTOEXEC.BAT file
A file of text commands that is placed on the disk drive that the computer boots from, and which sets up various items before programs are run.

Background
1 The screen that is visible outside the current active window.
2 An inactive window or an icon whose program can be working without attention from the keyboard or mouse. A program working in the background can be sorting or searching data or exchanging text or other files with another computer.

Binary file
A file of coded numbers that are meaningless when printed or displayed but which convey information. A program is always in binary-file form, but program-control files such as CONFIG.SYS, AUTOEXEC.BAT and WIN.INI are in ordinary readable text form.

Bitmap
A graphics image which is stored in the form of numbers that represent the intensity and colour of each part of the screen. A simple bitmap requires a lot of disk space, typically 100 Kbyte or more, for a screen. Other forms of file for graphics, such as PCX, compress this information considerably (if there are 500 consecutive red dots, for example, you need store only the information for one red dot along with the number of them).

Boot
To start up the computer either from a system disk or from a hard disk. The act of booting always checks and clears the memory.

Branch
A directory which is connected to the main (root) directory or which is a subdirectory or another directory.

Built-in font
A font which is permanently contained in a printer and which can be used by any software. The view of text on the screen will not necessarily correspond to the appearance when printed unless the screen can use an identical font.

Cartridge
A plug-in unit of memory used to alter the characteristics of a device. A cartridge font for a printer, for example, allows the printer to use fonts other than those built in.

Cascade
A set of windows which overlap but allow each title to be displayed so that it is possible to click on the top line of any one. Also applied to menus when one menu allows another to be opened with the first still visible.

Check box
A small square box that can contain an X or be blank, used in Windows to switch an option on or off.

Clicking
The action of quickly pressing and releasing the button (usually the left-hand button) on the mouse.

Clicking on name/icon
The action of placing a Windows cursor on a name or icon and then clicking the mouse button.

Clipboard
The temporary storage for text or graphics used by Windows to copy data from one application to another, or from one part of an application to another.

Close
To end the use of a window, either by double-clicking on the control-menu box, or by clicking on the control-menu box and selecting Close from the menu.

COM
1 Abbreviation for Communication used to indicate a serial port. The COM ports are numbered as COM1, COM2 etc.
2 An extension for a short type of program file.

Command button
The OK or Cancel word enclosed in a rectangular box and used to confirm or cancel a selection.

Communications settings
The settings of speed and other factors that are needed to make serial transfer of files possible.

CONFIG.SYS file
A file of text commands that imposes various settings on the computer before any programs, even MS-DOS itself, can be loaded. Changes to the CONFIG.SYS file have no effect until the machine is rebooted.

Confirmation message
A warning message that appears when you have chosen an action that might destroy files. You will be asked to confirm that you really intend to go ahead. Some confirmation messages can be turned off or restricted.

Control menu
The menu that is available for each Windows application, allowing you to move the window, minimize, close, expand etc.

Control-menu box
The small rectangle at the left of the Windows title bar which is used to bring up the control menu (single-click) or close the program (double-click).

Conventional memory
The first 640K of memory in which programs are run. All PC computers can be fitted with expanded memory (using an add-on board) and AT machines can also use extended memory (by adding more memory chips), but these additions are useful only if a program can recognize them. Windows can make use of extended memory, and there is little point in fitting expanded memory boards to an AT type of machine.

Copy
To place a copy of some selected portion of a screen on to the Clipboard for pasting into another program or file. This leaves the original unchanged, unlike the Cut action.

Ctrl-Alt-Del
The key combination that can be used to escape from a Windows program that appears to have locked up. Unless you alter the WIN.INI file, using Ctrl-Alt-Del while Windows 3.1 is running will not reboot the computer.

Cut
To select a piece of text or graphics and transfer it to the Clipboard, removing it from the current window. Compare *Copy*.

Default
A choice that is made for you, usually of the most likely option that will be needed. You need only confirm a default.

Desktop
The full screen on which all the windows, icons and menu boxes will appear as you make use of Windows.

Desktop pattern
A form of wallpaper which appears on the Windows desktop background so that you can distinguish the background more easily.

Dialog box

A box which contains messages, or which requires you to type an answer to a question that appears in the box.

Disc

A compact disc, also known as CD ROM, whose data tracks can be used for text, sound or graphics. Multi-media programs are distributed in this form, which is also used for collections of graphics images and for other large programs.

Disk

A magnetic disc, usually of the floppy type – the discs of a hard drive are called platters.

Double-clicking

The action used to select a program by placing the pointer over the program name and clicking the mouse button twice in rapid succession.

Dragging

The action of moving an object on screen by selecting it with the pointer, then holding the mouse button down and moving the mouse so as to move the object over the screen. The object is released when the mouse button is released. Some important dragging actions make use of auxiliary keys such as Shift, Ctrl or Alt.

Drag and drop

The action of dragging a file icon to another icon, such as the printer icon or a disk-drive or directory icon, and releasing the mouse button. When a file is dragged and dropped to the printer icon it will be printed (if it is printable and if the printer is on-line); when the file is dragged to a disk-drive icon it will be copied to that drive.

Embedding

The action of placing a drawing or an icon into a document, with the icon representing another document or drawing. A document dealing with the topic of using the mouse, for example, might have a drawing between two paragraphs on the screen. Clicking on that drawing would allow you to edit it using the program that created the drawing. See also *Linking*.

EMS

A standard for expanded memory that replaced the older EMS and which is compatible with another standard, the LIM 4.0 standard. EMS memory is not required for modern machines or programs.

Emulation

The imitation of another device, such as a laser printer emulating a dot-matrix printer, meaning that it will obey the control codes for the emulated machine. The word is also used for a form of memory management system that allows extended memory to be used as if it were expanded memory.

Enhanced mode

A method of using Windows, available only on 386 or 486 machines, which allows the disk to be used as part of memory, full use of extended memory, and the ability to run non-Windows applications inside windows. Also known as 386-Enhanced mode.

Exclusive application

A program running in a 386 or 486 machine which is using a full-size window and running at maximum speed because the processor is not sharing its time with any other program.

Expanded memory

A form of added memory which uses a plug-in board fitting one of the expansion slots of any PC machine, XT or AT. Some programs, but only some, can make use of this memory for data, such as for very large spreadsheet files. Expanded memory boards are unnecessary for modern AT386 or 486 machines, because some extended memory can be used as if it were expanded memory (using EMM386.EXE).

Expansion slot

The socket (usually one of 4 to 8) within the computer which will accommodate a plug-in card that enhances the capabilities of the machine. Such slots are used for video cards, disk controller, network card and other add-on devices.

Extended memory

The added memory of a modern PC machine, not possible on the old XT design. The motherboard generally provides for a machine to be supplied with at least 1 Mbyte of memory, and for more memory to be simply added directly (not in expansion slots). This memory can be used by only a few programs, Windows in particular. Machines in the 386 or 486 class can run software that allows some extended memory to be used like expanded memory for programs that require it.

Extension

The set of up to three letters following a full-stop (period) in a filename. For example, in the name MYFILE.TXT, TXT is the extension. The extension letters of a filename are used to indicate the type of file.

Flow control

A method of ensuring that serial data sent from one computer to another is synchronized, often by sending handshaking signals to indicate ready to send and ready to receive.

Font or fount

A design of alphabetic or numerical characters, available in different sizes and styles (Roman, Bold, Italic).

Footer

A piece of text that appears at the bottom of each printed page in a document.

Foreground

1 The part of the screen which contains the current active window.
2 The program which is currently under keyboard control and taking most of the processor time (see also *Background*).

Full-screen application

A program that can run only on the full screen, not in a Window. This applies to older programs running on XT or 286AT machines; a 386 or 486 machine can run any application inside a small window.

Graphics resolution

The measure of detail in a picture, in terms of dots per inch or dots per screen width. The higher the resolution the better the appearance, the longer it takes to print and the more memory it needs.

Handshake

See *Flow control*.

Header

A piece of text that appears at the top of each printed page in a document.

High Memory Area (HMA)

Part of the memory of a 386 or 486 machine which can be used by some files when Windows is installed.

Highlight
A method of marking an icon or text, using a different shading or colour.

Icon
A graphics image that represents a program or menu selection which can be used (made active) by clicking the mouse button over the icon.

Inactive window
A window which contains visible text or graphics but which is not currently being used by the mouse or keyboard.

Linking
A form of embedding in which the embedded document or picture retains links to the program that created it. If you click on an icon for a linked picture, for example, you can edit the picture, and the new edited version will affect any other document linked to that icon – changing one copy changes all (in fact, there is only one copy, used by all the documents in which it is linked).

LPT
An abbreviation of Line Printer, used to mean the parallel port (also indicated by PRN). When more than one parallel port is available, these will be numbered as LPT1, LPT2 etc.

Macro
A recorded file of a set of actions, allowing the actions to be repeated by replaying the file. In Windows, the Recorder is the method used to create and work with macros. Many programs, such as Word for Windows or Lotus 1-2-3, contain their own macro system

Mark
To select text or programs.

Maximize
The action of making a window expand to fill the screen. This can be done from the Control menu, or by clicking on the up-arrow at the right of the title bar.

Memory resident (or TSR)
A program which is loaded and remains in the memory of the computer rather than being run and discarded as most programs are. Such a program can be called into use by a key combination (a hot-key like Print Screen), or it can permanently affect the machine until it is switched off (like KEYB).

MIDI
An acronym of Musical Instrument Digital Interface, a system for allowing a computer to control electronic musical instruments. Windows 3.1 provides for such control by way of sound files, but only if a suitable sound card is added in an expansion slot of the computer, and the appropriate instruments connected.

Minimize
The action of shrinking a window and the program in it to an icon.

Modem
A device which converts computer signals into musical tones and vice-versa, allowing such signals to be transmitted along telephone lines.

Motherboard
The main board of a computer, into which expansion cards, maths co-processors and memory chips can be plugged. Many machines allow the whole motherboard to be swapped so that the machine can be upgraded (for example, from a 286 motherboard to a 386).

Mouse

The small trolley whose movement on the desk controls the movement of a pointer or other indicator on the screen. The use of the mouse is central to Windows.

MSD

The Microsoft System Diagnosis utility of Windows 3.1 which will prepare a report on your system on screen, on paper or as a text file.

Non-Windows application

A program written with no regard to use within Windows. Such a program can be run under Windows control, but with few of the facilities of Windows unless the machine uses the 386 or 486 processor.

OLE (Object Linking and Embedding)

See *Linking, Embedding, Packaging.*

Packaging

The use of an icon to represent a piece of text or a drawing so that it can be embedded or linked in another document. When the document is printed, the icon is printed, but double-clicking on the icon when the document is on screen will show the packaged material. Packaging allows a program that cannot be used directly for embedding or linking to have its files represented as icons in this way.

Parallel port

The connector used for printers which sends data signals along a set of cables, eight data signals at a time. Also called a Centronics port.

Parameter

A piece of information needed to complete a command. For example, a COPY command would need as parameters the name of the file to be copied and the destination to which it had to be copied.

Paste

To copy a piece of text or graphics from the Clipboard into a window.

PIF

Program Information File. A file of text that contains data needed if Windows is to be able to run a non-Windows application. Windows 3.1 can use a universal DEFAULT.PIF file for all but a few programs.

Pixel

A unit of screen display, a dot, whose brightness and/or colour can be controlled. Nothing smaller than one pixel can be displayed.

Platter

An aluminium disc coated with magnetic material and used within a hard drive for storing digital information.

Point size

A printer's unit of type size, equal to $\frac{1}{72}$ inch.

Pointer

The shape on the screen that moves as you move the mouse. Windows uses several different shapes of pointers to indicate that the pointer will have a different action when it is over a different part of a Window. Some programs that run under the control of Windows will use other pointer shapes in addition to these types.

Printer driver

A program that determines how the printer makes use of the codes that are sent from the computer – using the wrong printer driver will result in very strange printed output, because different printers use different methods. Many dot-matrix printers, however, use Epson codes, and many laser printers use either Hewlett-Packard Laserjet codes or the universal PostScript system (from Adobe Corp.)

Proportional font

A font in which the spacing between letters is varied according to the space needed by each letter.

Reboot

Restarting the computer either by using the Ctrl-Alt-Del keys (a soft reboot) or by pressing the Reset key (a hard reboot). Either will wipe all programs and data from the memory. The Ctrl-Alt-Del method used when Windows 3.1 is running can be arranged so as not to reboot the computer, but to restart Windows.

Restore button

The button that is placed at the right-hand side of the title bar when a window has been maximized – clicking on this button will restore the former size.

Screen font

A font that appears on the screen to indicate or simulate the font that has been selected for the printer.

Scroll bars

The bars at the right-hand side and bottom of a window. Dragging the button in the scroll bar performs the action of moving the window over the text or picture, allowing a different portion to be viewed.

Select

To choose an action by clicking its icon (another click needed to run it) or to mark text or graphics for cutting.

Serial port

The connector used for sending or receiving data one bit at a time. This is used mainly for connecting computers to each other, either directly or by way of a modem through telephone lines. A few printers require a serial port connection, many others allow it as an option. The serial ports are referred to by the letters COM.

Soft font

A font which is not built-in or in cartridge form, but sent as a file from the computer to a printer. Such a font needs to be loaded again after either the printer or the computer has been switched off. Such a font can be made to appear in identical forms both on screen and on paper.

Sound card

An add-on card that fits into an expansion slot allowing sound outputs to be taken to amplifiers and loudspeakers (some cards incorporate a small amount of amplification). Such a card, of which SoundBlaster is typical, allows sound effects to be incorporated into Windows actions.

Spool

To store printer information in memory so that it can be fed out to the printer while the computer gets on with other actions. If a lot of text needs to be spooled, it is useful to allow the spooling to take place in expanded or extended memory.

Standard mode

The normal way of using Windows on an AT machine, particularly the 286AT form of machine; see also *Enhanced mode*. Preferable even on an 80386 machine unless you require to run non-Windows programs within windows.

Swap file

Part of the hard disk used in enhanced mode to swap with memory so that the memory is not overloaded.

Text file

A file that contains only a limited selection of codes for the letters of the alphabet, digits and punctuation marks.

Tiling

An arrangement of windows in which there is no overlapping, unlike Cascade.

Title bar

The strip at the top of a Window that contains the title of the application, and also the control-box and minimize/maximize arrows.

TrueType font

A form of soft font packaged with Windows 3.1 which presents the same appearance on the screen as on paper, allowing you to be much more certain that what you see is what you eventually get.

UMB

A small amount of RAM which can use addresses in the 640 Kbyte to 1 Mbyte range, for storing small drivers and utilities. Accessible by using EMM386 on 80386 (and upwards) machines.

Vector font

A font that consists of instructions to draw lines, as distinct from a bit-font, which is a pattern of dots. A vector font can be easily scaled to any size.

Virtual machine

Referring to use of a 386 machine in which each application can be run in its own portion of memory.

Virtual memory

The use of a hard disk as if it were part of the memory of a 386 computer.

Windows application

A program that has been designed to run within Windows, and which will not run unless Windows is being used. All such programs present the same pattern of controls (the user interface) making them easier to learn.

Appendix B

Abbreviations and acronyms

ANSI
American National Standards Institute. The title is used for a number code system that follows the ASCII set for numbers 32 to 127, and specifies characters for the set 128 to 255.

ASCII
American Standard Code for Information Interchange. The number code for letters, numerals and punctuation marks that uses the numbers 32 to 127. Text files are normally ASCII or ANSI coded.

AT
Advanced Technology. The designation used by IBM in 1982 for the computer that succeeded the PC-XT.

BIOS
Basic Input Output System. The program in a ROM chip that allows the computer to make use of screen, disk and keyboard, and which can read in the operating system.

CAD
Computer Aided Design. A program that allows the computer to produce technical drawings to scale.

CD-ROM
A form of read-only memory. Consisting of a compact disc whose digital information can be read as a set of files.

CGA
Colour Graphics Adapter. The first IBM attempt to produce a video graphics card.

CISC
Complex Instruction Set Chip. A microprocessor which can act on any of a very large number (typically more than 300) instructions. All of the Intel microprocessors to date are of this type. See also *RISC*.

CMOS
Complementary Metal-Oxide Semiconductor. A form of chip construction that requires a very low current. As applied to memory, a chip that allows its contents to be retained by applying a low voltage at negligible current.

CP/M
Control, Program, Monitor. One of the first standard operating systems for small computers.

CPU
Central Processing Unit. The main microprocessor chip of a computer.

CRT
Cathode Ray Tube. The display device for monitors used with desktop machines.

CTS
Clear To Send. The companion handshake signal to RTS in the RS-232 system.

DCE
Data Communications Equipment. A device such as a computer that send out serial data along a line.

DIL
Dual In Line. A pin arrangement for chips that uses two sets of parallel pins.

DIP
Dual In-line Package. A set of miniature switches arranged in the same form of package as a DIL chip.

DOS
Disk Operating System. The programs that provides the commands that make a computer usable.

DSR
Data Set Ready. Another form of handshaking signals for RS-232.

DTE
Data Terminal Equipment. A receiver of serial data such as a modem.

DTR
Data Terminal Ready. The RS232 companion signals to DSR

DTP
Desktop Publishing. The use of a computer for composing type and graphics into book or newspaper pages.

EEMS
Enhanced Expanded Memory System. A standard for adding memory to PC/XT machines, not used on modern machines.

EGA
Enhanced Graphics Adapter. The improved form of graphics card introduced by IBM to replace CGA.

EISA
Enhanced Industry Standard Architecture. A system for connecting chips in a PC machine which allows faster signal interchange than the standard (ISA) method that has been used since the early PC/AT models.

EMS
Expanded Memory System. The original standard for adding memory to the PC/XT machine, now seldom used.

ESDI
Enhanced Small Device Interface. A standardized method of connecting hard disk drives to a computer, now superseded by IDE and SCSI.

GEM
Graphics Environmental Manager. An early GUI program.

LCD

Liquid Crystal Display. A form of shadow display which is used on calculators and portable computers. It depends on the action of materials to polarize light when an electrical voltage is applied.

LCS

Liquid Crystal Shutter. An array of LCD elements used to control light and so expose the light-sensitive drum in a laser printer. The LCD bar is used as an alternative to the use of a laser beam.

LED

Light Emitting Diode. A device used for warning lights, and also as a form of light source in laser-style printers.

LIM

Lotus-Intel-Microsoft. A standard for the use of expanded memory agreed by these three major companies.

MCA

Micro Channel Architecture. A system proposed and used by IBM as a way of connecting chips within a computer, intended to replace the AT-bus (ISA).

MDA

Monochrome Display Adapter. The first type of video card used in IBM PC machines.

MIDI

Musical Instrument Digital Interface. A standard form of serial data code used to allow electronic instruments to be controlled by a computer, or to link them with each other.

MS-DOS

Microsoft Disk Operating System. The standard operating system for the PC type of machine.

NTSC

National Television Standards Committee. The body that drew up the specification for the colour TV system used in the USA and Japan since 1952.

OCR

Optical Character Recognition. Software that can be used on a scanned image file to convert images of characters into ASCII codes.

OS/2

An operating system devised by IBM and intended to replace PC-DOS (the IBM version of MS-DOS).

PAL

Phase Alternating Line. The colour TV system devised by Telefunken in Germany and used throughout Europe apart from France.

PBX

Private Branch Exchange. Sometimes a problem for using modems.

PSS

Packet Switch Stream. A method of transmitting digital signals efficiently along telephone lines.

RAM

Random Access Memory. All memory is random access, but this acronym is used to mean read-write as distinct from read-only memory.

RGB

Red Green Blue, the three primary colour TV signals. A monitor described as RGB needs to be supplied with three separate colour signals, unlike a monitor that can use a composite signal.

RISC

Reduced Instruction Set Chip. A microprocessor that can work with only a few simple instructions, each of which can be completed very rapidly.

RLL

Run Length Limited. A form of high-density recording for hard disks.

ROM

Read-Only Memory. The form of non-volatile memory that is not erased when the power is switched off.

RS232

The old standard for serial communications.

RTS

Request to Send. A handshaking signal for RS-232.

SCART

The standard form of connector for video equipment, used on TV receivers and video recorders.

SCSI

Small Computer Systems Interface. A form of fast-acting disk drive interface which allows for almost unlimited expansion. Used mainly on Mac machines, but also found (in a less standardized form) for some PC devices.

SECAM

Sequence Colour à Memoire. The French colour TV system, also used in Eastern Europe and the countries of the former USSR.

SIMM

Single Inline Memory Module. A slim card carrying memory chips, used for inserting memory in units of 1 Mbyte or 4 Mbyte.

TIFF

Tagged Image File Format. One method of coding graphics images that is widely used by scanners.

TSR

Terminate and Stay Resident. A form of program that runs and remains in the memory to influence the computer.

TTL

Transistor-Transistor Logic. A family of digital chips. The name is often used to mean that a device will work on 0 and +5V levels.

VDU

Visual Display Unit. Another name for the monitor.

VEGA

Video Extended Graphics Association. A group of US manufacturers who have agreed on a common standard for high-resolution graphics cards.

VGA

Video Graphics Array. The video card introduced by IBM for their PS/2 range of computers.

Appendix C

Hexadecimal number scale

All computing depends on the use of number codes. Some of the numbers are used to refer to locations in memory, each of which is numbered, some are used to mean commands, characters (using ASCII code) and other references. Each of these numbers is internally a set of 1s and 0s. Binary code like this is fine for machines, because with only two possibilities to work with, the chances of the machine making a mistake become very remote. Humans, however, are not ideally suited to working in binary numbers without making mistakes, simply because the stream of 1s and 0s becomes confusing, and an obvious step is to use a more convenient number scale.

Just what is a more convenient number scale is quite another matter. Most people work with the ordinary 0 – 9 scale of denary numbers, based on counting in tens. Memory analysing programs like MSD, and memory management programs of all types, however, are written as much for the convenience of professional programmers, who use hexadecimal numbers, as for the ordinary computer user. Hexadecimal means scale of sixteen, and the reason that it is used so extensively is that it is naturally suited to representing binary bytes.

Four bits, half of a byte, will represent numbers which lie in the range 0 to 15 in our ordinary number scale. This is the range of one hex digit (Figure A.1). Since we don't have symbols for digits higher than 9, we have to use the letters A, B, C, D, E and F to supplement the digits 0 to 9 in the hex scale. The advantage is that a byte of data can be represented by a two-digit number, and a complete address by a five-digit number.

Converting between binary and hex is much simpler than converting between binary and denary. The number that we write as 10 (ten) in denary is written as 0A in hex, eleven as 0B, twelve as 0C and so on up to fifteen, which is 0F. The zero doesn't have to be written, but programmers get into the habit of writing a data byte with two digits and an address with four or more even if fewer digits are needed. The number that follows

Denary	Hex	Denary	Hex
0	00	8	08
1	01	9	09
2	02	10	0A
3	03	11	0B
4	04	12	0C
5	05	13	0D
6	06	14	0E
7	07	15	0F

Figure A.1 *Denary and hex digits for 0 to 15.*

0F is 10, sixteen in denary, and the scale then repeats to 1F, thirty-one, which is followed by 20. The maximum size of byte, 255 in denary, is FF in hex.

When we write hex numbers, it's usual to mark them in some way so that you don't confuse them with denary numbers. There's not much chance of confusing a number like 3E with a denary number, but a number like 26 might be hex or denary. The convention that is followed by many programmers is to use a capital H to mark a hex number, with the H sign placed after the number. Most of the MS-DOS memory utilities assume that you will type in hex numbers, and they will not work with anything else, and addresses in MSD make use of hex numbers also.

Before we can get much further with addresses and their contents, we need to look at the way that addresses are organized. Because the 8088 and 8086 chips were developed from the older 8-bit 8088 type, there has been a strong family resemblance, and one thing that has carried over is the storing of numbers in 16-bit units, one-word units, used also in later chips. Now one word of 16 bits can represent a scale of ten numbers from 0 to 65536, which is hex 0000H to FFFFH, a range of 64 Kbyte, but the construction of the 8088, 8086 and 80286 chips allows for the use of 20-bit numbers for addresses. This makes the hex range of numbers 00000 to FFFFF, 0 to 1048575.

The ordinary 0 – 640 Kbyte range of memory uses the numbers 00000 to 9FFFF, and the rest of the memory uses the numbers A0000 to FFFFF. These numbers are split into two four-digit groups, the segment number and the displacement number. All of the Intel processors from 8088 to 80286 make use of these two address numbers in the same way. An address is created by adding the numbers, but not in a straightforward way. The numbers are placed with the segment number shifted one place to the left, so that adding A000 to 0100 will give A0100, a five-digit number. For the 80386 and 80486 processors, the address can consist of a full 32-bit set such as 17FE2A55 in hex. This can be converted to the 5-digit form when the chip is being used with MS-DOS. The higher numbers could be used for virtual memory under a suitable operating system. The numbers that are shown for memory addresses by MEM and other diagnostic programs are the segment numbers, the first four hex digits of each memory number. An address of A000 therefore refers to the 64 Kbyte of memory from A0000 to AFFFF.

Appendix D

Some other books

The following books are concerned with aspects of computing that have been touched on in this book, but not in enough detail to satisfy your needed once you have started to use the computer in earnest. All of these books are of UK origin, so that their cost is lower, and problems that are encountered by UK users of US software are more likely to be discussed. Butterworth-Heinemann Step-by-Step books, which deal with the essence of topics in a simple way, are indicated by S/S.

Newcomers to computing should consult the new series of Made Simple books from Butterworth-Heinemann.

General computing

Newnes PC User's Pocket Book by Ian Sinclair (Butterworth-Heinemann)
Computer Science by Ian Sinclair (Butterworth-Heinemann)

Operating systems and Shells

Newnes MS-DOS Pocket Book (2nd. Ed.) by Ian Sinclair (Butterworth-Heinemann)
Newnes Windows 3 Pocket Book by Ian Sinclair (Butterworth-Heinemann)
MS-DOS 5.0 S/S by Alan Balfe (Butterworth-Heinemann)
Using Windows 3 S/S by Arthur Tennick (Butterworth-Heinemann)
Using MS-DOS 3.3 to 4.1 S/S by Alan Balfe (Butterworth-Heinemann)

Word Processors and DTP

Illustrated WordStar by Randall McMullan (David Fulton Publishers)
Illustrated WordPerfect by Helen Banner (David Fulton Publishers)
PagePlus 2.0 DTP Companion by Ian Sinclair (Sigma Press)
Using WordPerfect for Windows S/S by Arthur Tennick (Butterworth-Heinemann)
Using Locoscript PC (V1.5) S/S by John Campbell (Butterworth-Heinemann)
Using MS Word 5.0 S/S by Roger Carter (Butterworth-Heinemann)
Using Word for Windows S/S by Alan Balfe (Butterworth-Heinemann)
Using WordPerfect 5.0 S/S by Anne Gautier (Butterworth-Heinemann)
Using Wordstar 5.0, 5.5, 6.0 S/S by Alan Balfe (Butterworth-Heinemann)
Ventura 3.0/4.0 for Windows S/S by John Campbell (Butterworth-Heinemann)
PageMaker 4.1 for Windows S/S by Alan Balfe (Butterworth-Heinemann)
Student's Guide to Desktop Publishing by Ian Sinclair (Butterworth-Heinemann)

Spreadsheets

Quattro Pro 3 S/S by P.K. McBride (Butterworth-Heinemann)
Starting Lotus 1-2-3 by Ian Sinclair (Dabs Press)
Using Lotus 1-2-3 for Windows S/S by Arthur Tennick (Butterworth-Heinemann)
Student's Guide to Spreadsheets by Ian Sinclair (Butterworth-Heinemann)
Using Excel V.3.0 S/S by Roger Carter (Butterworth-Heinemann)
Using Lotus 1-2-3 Macros S/S by Ian Sinclair (Butterworth-Heinemann)
Using Lotus 1-2-3 Release 3 S/S by Stephen Morris (Butterworth-Heinemann)
Illustrated Supercalc by Randall McMullan (David Fulton Publishers)
Get Going with Microsoft Excel 3 by S.S. Khalsa (Sigma Press)

Databases

Clipper Database Programming by Mike Towle (Sigma Press)
Database Applications in Engineering by George Burns (Sigma Press)
Inside dBASE IV by Mike Lewis (Sigma Press)
Mastering Masterfile by Ian Sinclair (Sigma Press)
Paradox 3.5 for Windows S/S by P.K. McBride (Butterworth-Heinemann)
Using Superbase 2 & 4 S/S by Arthur Tennick (Butterworth-Heinemann)
Using dBASE IV S/S by Roger Carter (Butterworth-Heinemann)
Using Sage Retrieve III by Ian Sinclair (Sigma Press)
Student's Guide to Databases by Garry Marshall (Butterworth-Heinemann)
Mastering DataEase by Francis Botto (Sigma Press)
Power of Paradox by Francis Botto (Sigma Press)

Graphics and Drawing

Corel Draw S/S by John Campbell & Marion Pye (Butterworth-Heinemann)
Computer-Aided Design on a Shoestring by Ian Sinclair (BSP)
Illustrated AutoSketch 3 by Ian Sinclair (David Fulton Publishers)

Upgrading machines, hardware, and diagnosing problems

Newnes Data Communications Pocket Book by Michael Tooley (Butterworth-Heinemann)
PC Memory S/S by Ian Sinclair (Butterworth-Heinemann)
Newnes PC Printers Pocket Book by Stephen Morris (Butterworth-Heinemann)
Newnes Hard Disk Pocket Book by Allen & Kay (Butterworth-Heinemann)
Hard Disks S/S by Ian Sinclair (Butterworth-Heinemann)
Using Disk & RAM Utilities S/S by Ian Sinclair (Butterworth-Heinemann)

Pocket Boook of Upgrading Your PC by Steve Heath (Butterworth-Heinemann)

Newnes 8086 Family Pocket Book by Ian Sinclair (Butterworth-Heinemann)

Programming the PC machine

Newnes C Pocket Book by Conor Sexton (Butterworth-Heinemann)

Simple C: A Beginner's Guide by Ian Sinclair (David Fulton Publishers)

Modula 2 on Amstrad and Compatibles by Ian Sinclair (David Fulton Publishers)

Starting MS-DOS Assembler by Ian Sinclair (Sigma Press)

Using Quick BASIC 4.5 S/S by Stephen Morris (Butterworth-Heinemann)

Visual Basic S/S by Stephen Morris (Butterworth-Heinemann)

Programming in GW BASIC by P.K. McBride (Butterworth-Heinemann)

QBASIC Beginners Book by Ian Sinclair (Bruce Smith Books)

Visual Basic Beginners Book by Ian Sinclair (Bruce Smith Books)

Student's Guide to Program Design by Lesley Anne Robertson (Butterworth-Heinemann)

Student's Guide to Programming Languages by Malcolm Bull (Butterworth-Heinemann).

Index